COLD COMFORT

COLD
COMFORT
Keeping Warm in the Outdoors
GLENN RANDALL

LYONS & BURFORD, PUBLISHERS

Printed in the United States of America

10 9 8 7 6 5 4

Library of Congress Cataloging-in-Publication Data

Randall, Glenn, 1957-
Cold comfort.

Includes index.
1. Clothing, Cold weather. 2. Sport clothes.
I. Title.
TT649.R35   1987   646'.3'0911   87-26197
ISBN 0-941130-46-0

Typeset by Fisher Composition, Inc.

NOTE

Hollofil®, Neoprene®, Hypalon® and Teflon® are registered trademarks of the Du Pont Company. Quallofil and Thermolite are Du Pont certification marks for products meeting its quality standards. Polarguard® is a registered trademark of Reliance Products. Bion-II® is a registered trademark of Biotex Industries. Gore-tex® is a registered trademark of W.L. Gore & Associates. Entrant® is a registered trademark of Toray Industries.

CONTENTS

INTRODUCTION

I began writing this book on a January night in 1987 in Colorado's Rocky Mountain National Park. I was camped at 11,840 feet, at Chasm Lake, a frozen alpine tarn set far above timberline beneath the 1,600-foot east face of Longs Peak. A full moon had risen shortly after sunset and was painting the cirque's pale granite walls in sharp-edged tones of black and white. The temperature hovered at five degrees above zero Fahrenheit. A boisterous breeze rippled my clothes as I sat outside eating dinner. And yet I was comfortable: warm, well-fed and awed to be alone in winter in the most beautiful amphitheater in Rocky Mountain National Park.

Learning to deal with the cold has many rewards. There's a real pleasure in watching a white-tailed ptarmigan nibbling lunch in a 30-mph gale above timberline, seemingly unconcerned about the spindrift whistling past its bare toes. There's a pleasure, too, in skiing when the snow is trackless and deep, usually right at the storm's height, before the powder hounds can flatten every flake. And there's a raw pleasure in being able to laugh when you're butting head-to-head with the worst weather the mountains can fling at you.

The right gear was only one of the reasons I was comfortable at Chasm Lake. Another was the experience distilled from 15 years of climbing and skiing—"experience" defined as "what you get when you don't get what you want." During many of my mountain experiences, being warm and dry was what I wanted most of all. I've spent almost seven months, in total, making tough journeys through the arctic paradise of the Alaska Range, home to Mt. McKinley. Climatologists estimate that the climate only halfway to McKinley's summit is the equivalent of the North Pole's. Temperatures up high routinely hit 40 below even during May, the peak climbing season. I've also spent time in the dry, bitter winds that constantly scour the slopes of Argentina's 22,834-foot Aconcagua, highest peak in the Western Hemisphere. What I've learned in the mountains about gear, my body and how they interact has made it possible for me to enjoy being outside in almost any kind of frigid weather. Cold, in fact, is easier to deal with than extreme heat. You can always add more insulation, but there comes a point when you can't take anything else off.

The double-entendre in the book's title is intentional. I can't claim I'm as lazily unconcerned about the temperature when it's 40 below outside my tent as I am sipping beer on my back porch on a June evening. But the animal in me was comfortable enough at 40 below to feel keen enjoyment in being in the mountains. With the knowledge in this book, the right gear and a little bit of your own experience—"getting what you don't want"—you can do the same.

· 1 ·
SECOND SKINS:
Long Underwear and the Physiology of Sweat

"What is warm when wet? A hot tub!" wrote Jack Stephenson in his Warmlite Equipment catalog. Nothing else truly is. To stay warm in the cold, you must know water's insidious ways. Water is everywhere—in snow, rain, air, your breath, your sweat. The physics of water dictate the terms of cold-weather comfort in more ways, literally, than meet the eye.

Let's start with sweat. Sweat, obviously, is intended to cool you—and it does this with a vengeance. It takes about 973 BTUs (245 food calories) to turn one pint of liquid water at skin temperature into water vapor. One BTU (British Thermal Unit) is the amount of heat required to raise the temperature of one pound (one pint) of water by one degree Fahrenheit. That energy is sucked directly out of your body, a blessing if you're slogging across the Sahara, a potential disaster in the cold once you stop the heat-producing activity that generated the sweat. Your body produces sweat at rates ranging from nearly zero to quarts per hour. No one knows the exact limit. Let's say you sweat out just half a quart between breakfast and lunch, an easy thing to do if you're careless about heat regulation. Evaporation of that much sweat requires as much energy as is needed to raise the temperature of two gallons of water by 61 degrees F. Think of setting a two-gallon bucket of ice-cold water on your stomach and warming it to a comfortable lukewarm temperature of 93 degrees with body heat alone. The amount of heat lost becomes apparent.

Well, if you're overheating, what's wrong with sweating, even if it's cold? Several things. First, that sweat is only about 50 percent effective in cooling you compared to its effectiveness in evaporating from bare skin in the summertime. The reason is that much of that sweat vapor never escapes your clothing. The amount of moisture that air can hold depends on its temperature. The warm air next to your skin can hold lots of moisture. But as that warm, moist air heads toward the cooler outer layers of your clothing, much of the moisture condenses. It takes heat energy to turn liquid water on your skin into vapor, and that energy is released when the vapor condenses back into liquid water. The net result is ineffective cooling, more sweat produced and water trapped in your clothing.

The water not only ruins the effectiveness of some insulations, but also creates a tremendous "heat debt" that you have to pay as soon as you stop moving. The water will distribute itself through your clothing until it gets close enough to your skin for your body's warmth to force it to evaporate again. Your skin is also likely to be wet from sweat that hadn't evaporated at the time you quit moving. The amount of heat extracted by evaporation will be the same as when you were active, but your much-diminished heat output at rest will no longer be adequate to replace it. Soon you'll be shivering.

The first secret to staying warm, then, is staying cool—cool enough so that you sweat the minimum amount possible. That means stopping to adjust your clothing as soon as you start to overheat. Too many people overdress in the wintertime. It's as if they're afraid of getting cold, and so bundle up like a thin man in a Halloween fat costume. The most extreme example of this that I've heard of occurred in 1965. A soldier was admitted to the hospital in Fort Wainwright, Alaska, who had been chopping wood outside when it was 40 below. The diagnosis: heatstroke. The cause: too much clothing.

Avoiding sweating in the winter sounds simple, doesn't it? After all, it's *cold* out there. Unfortunately, your body's physiology dictates otherwise.

You may think you actively sweat only when your skin feels hot. You're only half right. True, when skin temperature rises above normal, a spinal reflex tells the sweat glands in your skin to get cranking. But you also start sweating when your core temperature goes up through physical activity. Heat production during maximum exertion can reach twenty times the amount produced when you're sitting still. That increase can only be sustained for a few minutes, but a five-fold increase can be sustained for hours. In either case, your hot muscles heat your blood. The temperature regulator in your hypothalamus, a section of your brain, then stimulates the sweat glands to get rid of that excess heat. Even if some areas of your skin are cold, other areas may perspire. Your back is often one such area because your pack, usually filled with extra clothing and a sleeping bag, adds a lot of insulation to that part of your body.

Another common sweat source is your head. For most of your body, the brain controls the diameter of blood vessels near the skin, opening them wide when there is heat to be disposed of, closing them down when the core needs every BTU it can get. The opening of the blood vessels is called vasodilatation. The closing down is called vasoconstriction. As I mentioned above, blood vessels near the skin in most of your body also respond to skin temperature, closing down when the skin is cold, opening up when it's warm. Blood flow at maximum is 100 times greater than at minimum. These mechanisms don't affect your head, however, which is why working hard in the cold sometimes causes the sweat of your brow to run into your eyes at the same time your hands feel like numb claws. Your core is warm, and that warm blood is circulating vigorously through your head, causing the skin to sweat. However, the blood vessels in your hands, if exposed to cold enough air, can be constricting while your head is sweating, so your hands feel cold. This phenomenon is why you sometimes see marathon runners wearing shorts, a t-shirt and gloves. It's also why your head and neck can provide the escape hatch for as much as 25 to 50 percent of your total heat loss.

The old saying, "If your feet are cold, put on a hat," is based on these facts as well. Your feet are a long way from your core, which means blood is already cooling when it gets there. Your feet also have a larger surface area in relation to their volume than your torso. That means a proportionally greater area from which to lose heat. And your feet are well supplied with sweat glands, with about 4,000 per square inch compared to 640 per square inch on your back. Evaporative heat loss can be high. If you start to get a little chilly, your brain restricts blood flow to your feet and they start to get cold. Putting on a hat helps stem that overall heat loss. Your body now has some warm blood to spare, so it responds by sending some out to your nippy toes. Usually that's enough to warm them up, at least eventually. Your hands get cold easily for the same reasons that your toes do. Wearing a hat can help keep them warm, too.

Avoiding sweating entirely, even in a cold environment, is nearly impossible, though you should do everything you can to minimize the problem. To deal with what sweat is produced, you should wear clothing next to your skin that dries fast and retains its ability to trap air even when damp.

COTTON

I've learned the hard way (several times, I have to confess) that cotton is emphatically the wrong material for cold-weather use. The most recent occasion when I got that lesson frozen into my skin was during an hour-long spring training ride on my bicycle. The sun, which had been beating down when I started, had tempted me to leave the house wearing nothing on my back but a cotton t-shirt. I had noticed the black cloud looming up over the foothill peaks just west of my home in Boulder, Colorado, but with cocky nonchalance, I thought I would be home before the storm hit.

I wasn't. I was still eight miles from safety when a ferocious rainstorm broke loose, accompanied by gale-force winds. I was soon forced to stand up in my lowest gear to make progress on level ground, and even then I could only travel at a walking pace. The pelting rain immediately plastered the t-shirt to my skin. I became so miserable that when a passing driver asked if I wanted a lift, I gratefully accepted.

Cotton sucks up water like a thirsty camel in the desert. In textile argot, its "moisture regain" is about 15 percent, making it one of the most absorbent of all fibers. That means that a completely dry sample placed in a chamber at 70 degrees with 95 percent relative humidity will absorb 15 percent of its weight in water. Textile people use moisture regain to assess a fabric's affinity for water rather than how much it soaks up when placed in a bucket because fabrics hold liquid water in two ways—by actual absorption into the fiber and by surface tension.

All fabrics hold water by surface tension. Cotton, however, also actually absorbs water. The water molecules are chemically held by "hydrogen bonds," to use the chemist's term. The absorbed water then acts as a plasticizer on the long cellulose molecules that make up cotton. The result: the cotton molecule loses all its resiliency. A cotton fiber is six times stiffer when dry than when wet. Wet cotton fabrics collapse and cling to your skin. That forces all the water the fabric is holding to evaporate directly off the skin, extracting the maximum possible amount of heat. In addition, wet cotton conducts heat much faster than dry cotton. The increase varies, depending on the degree of saturation, but as a point of reference, pure water conducts heat ten times faster than pure cotton, 20 times faster than dry air. A dry cotton t-shirt is actually mostly air.

Cotton also dries very slowly, both because it holds so much water, and because it takes more energy to break a hydrogen bond holding a water molecule to the fiber than it takes to drive off a water molecule that is simply sitting on the fiber. Cotton has no place in my winter outdoor wardrobe, and it shouldn't in yours.

WOOL

Wool is even more absorbent than cotton, with about a 29 percent moisture regain. Some textile textbooks call that an advantage because you feel dryer longer. However, a typical wool underwear top only weighs eight or nine ounces. You can sweat enough to saturate the garment in an hour or two if you ignore an overheating problem.

Another supposed advantage of wool is that the condensation of water vapor as it is absorbed into a dry sweater actually releases heat. That could happen when a sweater is taken from a warm, dry house into the outdoors on a chilly day, for example. However, this reaction only occurs when water *vapor* (not liquid like sweat or rain) is absorbed into the fiber. It still costs energy to drive that water back out again.

Wool does resist collapse when wetted, so it still provides some insulation. Generations of backwoods moms and dads have raised their kids to wear wool in the wintertime. It still works. But, like cotton, wool dries very slowly. I got a graphic demonstration of that in 1978, during my first expedition to Alaska. After a week of rain and wet snow, every garment we owned was soaked. When the sun finally came out, we festooned every ski, pole and ice axe with sodden garments. Among them was a knitted synthetic sweater, fuzzy inside, smooth outside. It was one of the first nylon pile garments I'd ever seen. Within an hour the pile sweater was warm and dry. At sundown the wool sweater was still damp. I put on the pile and jammed the wool sweater into a stuff sack, where it remained, useless, for the rest of the 35-day trip. Wool also makes many people itch like they've got the chicken pox (I get a rash if I wear it when I'm sweating) and it's considerably less abrasion-resistant than the synthetics.

A few companies offer winter underwear made of silk. Although it feels wonderful next to the skin, silk has a moisture regain comparable to wool. It dries slowly and is quite expensive.

SYNTHETICS

The best winter underwear is knitted from fibers made of polypropylene, polyester or polyvinyl chloride. None of these fibers absorbs water. All retain their full resiliency when wet.

Polypropylene was the first synthetic to become popular as winter underwear. Back in the early 1960s, some oil refineries were burning propylene (the building block for polypropylene), because they couldn't find enough companies willing to buy it. Lifa, headquartered in Norway, was probably the first company to make polypro underwear. Their stuff was nothing romantic. It didn't go to Everest for its trial run. Instead, it was first used in babies' diapers.

The first polypro underwear to arrive in the United States was very thin. A myth began to spread that "polypropylene wasn't warm in itself" and that it had to be used beneath other underwear. In truth, like all synthetics, polypropylene's warmth is proportional to its thickness, as people realized when thicker versions became available.

Polypropylene has a moisture regain of only .05 percent, which means it dries very fast. Wet polypro doesn't cling to your skin like wet cotton because it retains its resiliency when wet.

Polypropylene can be treated so it wicks. Water will then flow along the fiber from wet areas to dry ones. Water creeping up a small-diameter glass tube is an example of the same principle. Wicking's advantage is that it subtracts moisture from your skin without subtracting heat the way evaporation does. The moisture passes out into the outer layers of your clothing, where, at least in some conditions, some of the heat for evaporating it can come from the environment. Wicking also makes the fabric next to your skin dry faster by spreading the moisture over a larger area where more air can get to it. Visualize how long it takes for a glass of water to evaporate. Now picture how long it would take if you spread the same amount of water out on a sidewalk.

Polypropylene does have drawbacks. One is its low melting point, which means it must be washed in cold water and line-dried to avoid shrinkage. Wash it hot and you'll have doll clothes. Another is its tendency to amplify body odors. Sources at Du Pont and Patagonia say that polypropylene is oleophilic ("oil-loving") which means it latches onto and holds the oils from your skin, which in turn begin to reek. In my experience, polypropylene does indeed smell worse than other synthetic fibers after a few days in the field. They all begin to stink eventually, however.

Several companies now offer different versions of polyester long underwear. Polyester has a moisture regain of less than one percent and doesn't lose its resiliency when wet, so it meets the two fundamental criteria for winter underwear. Some of the new versions have a better "hand," (textile lingo for feeling good on your skin) than polypropylene, which can develop a rather plastic feel. Polyester is inherently more stable when heated than polypro, which means you can throw it in the washer and dryer. Polyester underwear also doesn't magnify body odors like polypro does.

A solid block of polypropylene is about one-third

lighter than a solid block of polyester of the same size, which should make polypro garments about one-third lighter. However, according to Mike Frankosky, a research scientist at Du Pont, the difference in the finished fabrics' weight is not quite that great, because polypro fibers in general don't hold a crimp quite as well as polyester, which means they must be made a little thicker to get a bulky yarn. Hollow-fiber polyester underwear is now available, which further reduces polypro's weight advantage.

Polyvinyl chloride is also used in a few brands of underwear. Like polypropylene, PVC is almost completely non-absorbent. Also, according to an article published in the Textile Research Journal, researchers at the *Institute de France* found that pure PVC also has the unusual property of generating a negative static-electric charge when it rubs against your skin. Most other fibers generate a positive charge. According to the article, a researcher named Deniker found that wearing a PVC garment for several days provided "considerable improvement" in the pain of rheumatism sufferers.

On the negative side, PVC wicks slowly. That doesn't mean it's bad stuff. Wicking's value is not always easy to detect in the field. Simply having underwear that won't absorb water and won't collapse when wet gets you 90 percent of what you want. Having it wick as well is just the final step.

All long johns tend to bind, particularly when worn under a knitted or woven pant. To reduce the friction of skin against underwear, try wearing a pair of nylon-Lycra tights against your skin. To reduce the friction of underwear on pants, add small patches of smooth nylon fabric to the inside of the pants' knee area.

No one has yet found a way to freeze-dry a hot tub. Until someone does, you'll only be comfortable if you stay dry by avoiding perspiring and by wearing synthetic underwear. Stay dry, and you'll find that staying warm is no sweat.

· 2 ·
FLORIDA IN THE ARCTIC:
The Surprise of
Vapor-Barrier Clothing

Ounce for ounce, the warmest clothing you can buy is also the least understood. I'm talking about vapor-barrier shirts, pants, socks and sleeping bag liners. These garments are made of waterproof fabric and they're designed to be worn near your skin.

A suit of vapor-barrier clothes can stop fully 15 percent of your normal heat loss, yet shirt, pants and socks together weigh only 7½ ounces, about the same as one medium-weight polypro turtleneck. VB clothes have an additional virtue: they warn you the moment you start to overheat. You can then stop and shed a layer before sweat can soak your clothes. If you do sweat a little even after shedding, wearing VB clothing lets you stop moving without instantly developing the shivers.

"But don't you sweat to death?" most people ask. Not necessarily. As the first chapter explained, you actively sweat only when a portion of your skin overheats, or when your core overheats, or both. You can keep active sweating to a minimum by carefully regulating the amount of insulation you're wearing. My friends and I routinely stop to strip down 15 minutes after we start up from any extended rest. We make other clothing-adjustment stops as necessary. Don't worry about holding up the rest of the team. You'll slow them down much more (besides endangering yourself) if evaporation from your sweat-saturated clothing chills you severely when you finally do stop.

Whether you wear a vapor barrier or not, sweating will extract its price. Without a vapor barrier, you pay it by being unpleasantly hot when you're moving, by losing excessive amounts of fluid, and by developing the shivers when you stop. With a vapor barrier, sweating without any evaporative cooling soon becomes so uncomfortable that you stop and shed clothing, as you should, rather than slogging on in your own version of Bogart wading chest-deep down the river in *The African Queen*.

Here's how vapor barriers work. Your body is about 57 percent water by weight. Much of that water lies between cells, not inside them. Because your skin is not like a plastic bag, some of that water is constantly evaporating—and when you're fighting to stay warm, that hurts. This water loss, called insensible perspiration because you don't sense its presence, goes on night and day, with no nervous system control, at the rate of about half a quart every 24 hours. Roughly 15 percent of the total heat lost by a sedentary person escapes through evaporation of insensible perspiration. VB shirts, pants and socks stop almost all of that evaporative heat loss.

Vapor-barrier clothing does take some getting used

to. Listen to one world-traveling mountaineer, Yvon Chouinard, describe his experiences.

"I experimented with vapor-barrier shirts in 1974 but dropped the idea because I didn't like it. I didn't understand the concept too well, and I found I couldn't regulate the warmth very much. I found that most of the time I was too sweaty. I had to keep taking things off. Where I really got sold on it was fly fishing in late October in Yellowstone for brown trout. I was wearing chest waders, pile pants, long underwear, pile booties, lightweight polypropylene, heavy polypropylene, pile jacket, wool shirt, waterproof anorak and a wool hat. I was standing all day long in 40-degree water up to my waist. The wind was blowing, it was snowing, the temperature was near freezing. I wasn't moving enough to stay warm and all my heat was being drawn away by this cold water. I could not stay warm, no matter how thick an insulation I used. Finally I used a vapor-barrier shirt as well. I found I could cut down on the insulation and still stay warm. I've used it on every expedition since."

Doing something outdoors in the cold that demands little exertion is the best place to start experimenting with VB clothing. Wearing VB clothing while ice-fishing, sitting in a duck blind, or just cooking a meal, demonstrates the strength of the clothing. Few people overdress so extravagantly that they actively sweat when standing still. Everyone loses insensible perspiration even when they're inactive, however. VB clothing stops that heat loss. It also keeps your outer insulation dry and prevents condensation on the inside of waterproof rain gear, further adding to your warmth.

Used correctly, VB clothing can also be valuable in more strenuous sports. First, be sure to shed clothing when you first start to overheat. Unzip the VB shirt at the neck, open the cuffs and unzip the VB pants at the sides if they're equipped with such zippers. If you're not wearing a pack, pull the shirt tails out of your pants. This will help promote ventilation through the "chimney effect": cool air enters at the bottom, is warmed by your body, and rises inside the shirt, absorbing moisture as it goes, before escaping at the neck. The shirt "draws" just like a chimney. When you stop, batten down the hatches to halt the instant chill-down so common when you quit moving in the winter.

Since it's very difficult to avoid at least a little sweating when working hard, you'll probably trap a little moisture inside the VB garments. Don't worry about it. Leave the VB garments closed up when you're resting or puttering around camp. Later, when you move on and start putting out lots of heat again, or when you snuggle into your bag and have lots more insulation, you can open up the VB clothing and finish drying out.

Many people find that wearing a VB garment next to their skin feels rather odd. I recommend wearing a thin, non-absorbent layer of underwear between your skin and the vapor barrier. That lengthens the time between the onset of sweating and when you first notice the sweat, meaning more sweat gets produced, but seems to be more comfortable. That system has the added advantage of reducing friction between your clothing layers, providing freer movement. Try it both ways and see what you like best.

Vapor-barrier clothing presents more advantages and fewer drawbacks as the temperature plummets. When I first bought a VB shirt I put it on next to my skin, added one thin layer on top and went for a bike ride on a windy, 40-degree day. To my astonishment, I was simply too hot. For active sports, I'd recommend VB clothing only for temperatures below freezing. It really comes into its own on days when water bottles freeze solid before noon—days when the high might be zero and the low 20 or 30 below. For inactive sports, some people recommend its use in temperatures below 65. Personally, I do quite well with conventional clothing down into the twenties, even when I'm inactive.

My first experiment with VB socks surprised me as much as that ride with the VB shirt. I put a plastic bag over a liner sock, then added a heavy wool sock and put the whole thing into a double mountaineering boot with a foam inner bootie and stiff outer plastic shell. On the other foot I wore the same layers without the plastic bag. Then I went for a hike in November in Rocky Mountain National Park. It proved to be fairly warm for late fall, with temperatures in the 40s. Two miles into the hike, I had to stop and remove the plastic bag. That foot was almost painfully hot.

VB socks have proved their worth many times over on colder days. I wore the same boots and socks, plus the vapor-barrier socks, when it was 40 below at 17,000 feet on McKinley. My toes grew cold, but didn't freeze.

When wearing VB socks, be sure to give your feet a chance to air out. My nightly procedure on McKinley was to remove the damp liner socks and pour a little athlete's foot powder onto my toes. The powder not only felt great, it helped dry the skin and prevented my toes from sprouting mushrooms. The foot powder's strong perfume also prevented my tent mates (and me!) from gagging on the odor of feet unwashed for a month straight. Then I put on spare liner socks that I had dried the previous night inside my clothing while I was asleep. The wet socks I had just removed went down inside my bibs to be dried in their turn. During the night the VB socks remained outside the sleeping bag. In the morning I turned them inside out, brushed off any frost, and put them on. Leaving your VB socks on for several days straight turns your feet into prunes, which makes them quite susceptible to blistering and a serious cold injury called trenchfoot, which I'll discuss fully later. People with hyperhidrosis (excessive sweating of the hands and feet) are particularly susceptible to these problems. About ten percent of the population has this condition. These people may find that their feet get so wet so quickly that VB socks do more harm than good.

Sleep is another sedentary activity where the VB principle works superbly. Using a VB sleeping-bag liner lets most people sleep comfortably in temperatures ten to 15 degrees colder than they could tolerate in their sleeping bag alone. Using a VB liner is particularly important when using a down bag because down loses loft when it gets wet. I usually leave my VB liner open at the neck. That allows some moisture to escape from my clothes while I'm asleep without that moisture condensing inside my down bag. If I'm fighting for every bit of warmth I can get, then I cinch the VB liner down tightly.

Coated-fabric vapor-barriers don't stop moisture penetration completely. On a couple of trips I've been so fanatic about keeping my down dry that I used a giant plastic bag over my cloth VB liner. Condensation appeared inside the plastic bag, meaning sweat vapor had penetrated the coated fabric and been stopped by the plastic. It's still important to dry your sleeping bag whenever possible, even when using a VB liner.

· 3 ·
THE WELL-DRESSED ONION:
Layering for Versatile Warmth

My sister calls it her "goose:" a plump, warm, goose-down jacket. She took it and one thin sweater when we went skiing on a blustery February day near Berthoud Pass, Colorado. In total, she easily had enough insulation; but the disadvantage of its form soon became apparent. Fifteen minutes into the tour, she started cooking inside her goose and had to stow it. Without it, however, the wind soon gave her a chill. On went the goose again, quickly followed by another cycle of overheating.

Amy's goose is a great garment for winter birdwatching, but not so good for a one-day ski trip. No one would willingly start a hilly bike ride on a machine with only one or two gears. Similarly, no one should go into the winter wilderness with just one or two garments, no matter how heavy.

The layers of clothes you bring in the winter should be like the corn-cob cluster of tightly spaced gears bike racers use. They want a very close match between their maximum sustained output of energy and the particular gear best suited to the terrain. To achieve this, each gear differs only a little from the adjacent one. In a similar fashion, you should bring many thin layers of clothing when you venture into the cold. That way you can accurately fine-tune the amount of insulation you're wearing to correspond to your heat output and the vagaries of temperature, altitude and wind.

I've talked already about the need to shed as soon as you start to overheat. It's just as important to don clothing the minute you start getting chilly. The earliest signs of dropping core temperature appear some time before the first shiver. Cold finger tips can result just from handling something cold, but if your entire hand is cold, think about adding a layer. One of the first signs of dropping core temperature for me is a stiffness in my wrist, a sign that my brain is beginning to shut down blood flow in my lower arm. I test for another sign by bending my elbow as far as it will go. If the fabric next to my skin feels cold when pressed against the skin, I know some blood-flow cutback is starting to occur. To stay comfortable in the cold, you need to become sensitive to small changes in your overall thermal comfort and to respond to them quickly, before your core temperature strays too far from normal.

An incident that happened to a client I was guiding on McKinley illustrates a common pattern perfectly. We'd been traveling steadily during the day, moving camp from 9,600 feet to 11,000 feet on the West Buttress. Ginny had no winter camping experience at all before the trip. The sun had warmed us during the day, but it had dropped behind the ridge as we began climbing the final mile into camp. Ginny had probably started getting a little chilly when the sun set, but hadn't wanted to slow everyone down by making another stop just a short ways from the camp.

When we got into camp, we dropped our packs and started building snow walls to shelter the tents from the persistent winds. Ginny neglected to add clothing to compensate for her diminished heat production now that she had stopped moving. Fortunately, she had the presence of mind to let me know that she was getting very cold. I threw up a tent, got her inside and into her sleeping bag. Then I cranked up a stove and started plying her with hot soup. Within half an hour she was feeling good again. But the discomfort was needless. If she'd simply put on another thin layer a mile from camp, then her parka as soon as she stopped, she would have been as comfortable as a baby in bathwater the whole evening.

THE CLOTHING SYSTEM

When you're wearing a lot of layers, it's important to think in terms of a system. Pile up all the clothing you need for the severe cold of the Alaska Range or Yellowstone in winter, for example, and it looks like it will fill a small trash can. Your outer layers have to fit over a tremendous amount of bulk, which often means going up a full size, particularly in parkas and shell gear. In fact, if you want your shell jacket to fit over your parka (not a bad idea if you've got a parka that doesn't have a waterproof shell), it may have to be two sizes larger than normal. The only way to tell is to try on every garment simultaneously before the trip.

If it's a major expedition you're planning, you need

to get your gear together months in advance so you can test it on several overnight trips. The little annoyances you'll find during such dry runs (or wet runs, if the gear's really faulty!) will become major hassles after a week of continuous cold. One of the most important lessons I've learned from sustained cold-weather expeditions is that the details make a big difference in comfort.

Every zipper, for example, whether on your clothing, your pack, your tent or your sleeping bag, should have a loop of string tied through the pull tab on the slider so you can grab it with gloves on. Do your pants tend to work their way down when carrying a heavy pack? Think about buying suspenders. Then try them out too. It may be that the buckles will end up underneath the pack waist belt and chafe.

Do you have enough pockets inside your clothes? You'll find you want to keep many little things handy or warm: sun block for your skin, a different sun block for your lips, a cigarette lighter or two, a pocket knife, perhaps film and an extra battery for your camera. Consider adding an extra pocket or two. I've got four now on my bibs, which look just like a farmer's bib overalls except they're made of synthetic pile. They came with one.

How much trouble is it to get into those pockets? Pockets closed with hook-and-loop (the most common trade name is Velcro) are a pain in the fingers in the winter. The hook part scratches your fingers if they're bare or latches onto your gloves if they're not. I cut the hook-and-loop out of the pocket on my bibs and replaced it with a light zipper.

Snaps provide another source of frustration for numb, begloved fingers. I've seen snaps on gaiters freeze so hard that it took ten minutes of prying with a Swiss Army knife's screwdriver blade while simultaneously thawing with a lighter to get them open.

As part of your gear prep, carefully check every item of used equipment. Think about how many times a year you actually use your winter paraphernalia. Let's say you're pretty enthusiastic and average one day a week for five months. That's about 20 or 22 days of use in a whole year. You can give your clothing that much use, the equivalent of a full year, in just three weeks during a big trip. Inspect each item you plan to bring and ask yourself if it can really endure another year of normal use. Check each seam of each garment by turning it inside out and trying gently to pull the seam apart. Are there places where the thread has worn through or some careless seamstress didn't backstitch a hem? Fix them.

If you can't fix them, or you don't think the fabric itself has enough life left, buy new—then go test it. There's nothing more embarrassing than landing on the glacier with brand-new equipment and finding that the gear had some hidden defect which causes it to fail at the first test. On my second trip to Alaska, our pilot dropped us off, without a radio, beneath the 6,000-foot south face of Mt. Hunter. One of my companions cinched down the spindrift collar on his shiny new pack and promptly ripped the drawstring completely out of its sleeve. My other companion discovered that the waist belt buckle on his new pack could simply slide off, which it had apparently done somewhere in transit. A bit wiser after that experience, I turned two new pair of gloves and a new pair of overmitts inside out before embarking on a big trip a few years later. All four of the gloves had a place where the lining had not been stitched down properly. The seam at the end of one overmitt liner had a gap where the sewing machine had simply missed the fabric.

When you think about your clothing system, think about ventilation. If it's easy to vent excess heat without stopping, you're more likely to do it than if you have to stop, doff your pack, strip, load the pack, put it back on and readjust the pack harness. Easy ventilation means zippers in strategic locations, like down your chest, under your arms and down the sides of your legs. I've even added 12-inch chest zippers to my long underwear tops, just to make ventilation easier. If you're wearing pants, (rather than bibs) try to get them with double-pull zippers so you can unzip from the ankle up to put them on, then unzip them from the waist down to ventilate them. Better yet, get separating double-pull zippers so you can unzip from the ankle up, then separate the zipper halves at the waist. That allows you to get your pants on or off even while wearing skis, snowshoes or crampons. Conventional knitted or woven pants rarely feature this kind of zipper, but they're common on good pile pants.

Pants work fine in moderate cold, but when the temperature plummets, they take second place to bibs for several reasons. First, cold air can rush into that annoying gap that appears between your pants and jacket when you bend over. Second, the belt frequently required to keep pants up can chafe under a pack or sled-pulling harness. Third, that belt also prevents air from flowing from your hard-working and therefore warm legs to your cooler torso. Wearing a pair of bibs, or very high-waisted pants with suspenders, lets that warm air circulate throughout your clothing. Bibs also close the gap at your waist and prevent chafing by a belt.

If you choose bibs, make sure they have a "dumpster": some kind of crotch zipper or zippered drop seat
to let you relieve yourself easily. Lack of a dumpster
forces you to remove every layer outside your bibs
whenever you have to go. In stormy weather that usually includes your shell jacket. Removing that lets snow
and wind in, and a lot of heat out.

Some people prefer knickers, either in bib or pant
form. Cinched down snugly below the knee, they do
offer great freedom of movement. Good ones can be
hard to find, however. Wearing knickers means you
have to buy knicker socks as well as gaiters designed to
close securely over the end of the knicker. It also
means you have to work the knicker sock up and over
the legs of your long johns. You can get some of knickers' freedom of movement in your pants or bibs by the
trick I mentioned before: add a patch of nylon to the
knee area where it binds against your long johns. Or
just pull up your pant leg a little before cinching down
the top of your gaiter. The extra pant fabric around
your knee gives you greater freedom of movement.

Climbers in extremely cold environments sometimes
use one-piece, insulated suits. Warm air can circulate
freely inside the suit, and no heat can escape at the
waist, so one-piece suits are very comfortable when it's
bitterly cold. One-piece suits lack versatility, however.
Too often you find yourself wanting your shell jacket,
but not your shell pants. Not many people really need
such specialized clothing, so one-piece suits are expensive and hard to find.

HEAT LOSS AND INSULATION

No clothing (with the exception of electric socks) actually generates heat. All clothing can do is slow the loss

of heat generated by your body. Besides evaporation, heat can be lost in three ways: conduction, convection and radiation. If you understand these processes, you'll have a better understanding of how different kinds of insulators work. That helps you build a clothing system that suits your needs, and helps you avoid getting taken by some salesman's hype. So take a deep breath, hold on to your pocketbook, and let's dive in.

Brrr! It's cold in here! That's because the heat from your body is conducting into this cold water. Conduction is the process in which the energy of hot molecules is transmitted by collision to cooler, less energetic molecules. It can account for more than half of the total heat lost through clothing. In knitted and woven fabrics, conduction through the clothing fibers themselves can account for 50 to 80 percent of the total conductivity. That's because the fibers take up ten to 40 percent of the fabric's volume and because the conductivity of the fiber is about ten times the conductivity of air.

In high-loft insulators, on the other hand, such as down and the polyester fiberfills, the insulator takes up only about one percent of the volume. Those fibers don't run directly from one surface to the other like little pillars holding up a roof. Instead, most run at an angle to the shell fabrics, which slows conductive heat loss through the fiber still further. Even a steel wool batting, with highly conductive fibers, is only about 12 percent less insulating than a wool one of equivalent thickness, according to the Army's Research Institute of Environmental Medicine in Natick, Massachusetts.

Although the conductivity of air is much lower than the conductivity of polyester, conduction through air itself is significant because air takes up much more space inside the insulator. Air's conductivity is not worth worrying about, however, because there's nothing you can do about it short of wearing a steel vacuum bottle as a shirt.

Climb out of the pond now, towel off before the evaporation chills you, and grab a cup of hot tea. See the steam rising off the cup? That's an example of spontaneous convection. The warm, steamy air just above the cup is less dense than the cold air. Air heated by a warm object tends to float upward. Cool air then replaces the warm air, to be warmed and float upward in its turn. A convection current is born.

Spontaneous convection was once thought to be a major avenue of heat loss in clothing. More recent work at Canada's Defense Research Establishment and Natick, however, has shown that convection plays almost no role in heat transfer through battings with a fiber content as low as .2 percent by volume. A batting with that few fibers is practically transparent. Convection can play a significant role in heat transfer through air when no fibers are interposed.

Convection doesn't have to be spontaneous; it can also be forced. Your movements, or high winds pummeling your clothing, can squeeze warm air out of gaps in your clothing at your waist, neck, wrists and ankles. Cooler air is sucked back in, resulting in a net loss of heat. If you're not wearing a windproof shell garment, wind can also force tiny air pockets between the fibers of your outermost garment to become convective when they otherwise would not. If, in addition, you're wet, watch out! Wind also increases the rate of evaporation. A wet person without windproof clothing on a cold, windy day is in real trouble.

Both wind and movement also disrupt the "boundary layer" of air around your clothing, which contributes some warmth. This "boundary layer" is not some specific quantity of air held next to your clothing; instead, it is a convenient way of referring to the fact that a warm object placed in cool, still air does not lose heat instantly. Convection, conduction and radiation all take time. The insulation value of something is nothing more than its ability to slow the rate of heat transfer. Insulation scientists therefore ascribe the rate of heat loss from an object in still air to the insulation value of the "boundary layer" of air. Since wind increases the rate of both conductive and convective heat loss, scientists say the "boundary layer" has been removed by the wind.

Heat loss through forced convection is the origin of the "wind-chill factor": the apparent drop in temperature caused by wind. Paul Siple came up with the

Wind Chill Chart														
Wind	*Temperature (Fahrenheit/Centigrade)*													
(Miles per hour)														
(Kilometers per hour)														
CALM	35	30	25	20	15	10	5	0	−5	−10	−15	−20	−25	−30
CALM	*27*	*−1*	*−4*	*−7*	*−9*	*−12*	*−15*	*−18*	*−21*	*−23*	*−26*	*−29*	*−32*	*−34*
	Equivalent Temperature (Fahrenheit/Centigrade)													
5 MPH	33	27	21	16	12	7	1	−6	−11	−15	−20	−26	−31	−35
8 KPH	*1*	*−3*	*−6*	*−9*	*−11*	*−14*	*−17*	*−21*	*−24*	*−26*	*−29*	*−32*	*−35*	*−37*
10 MPH	21	16	9	2	−2	−9	−15	−22	−27	−31	−38	−45	−52	−58
16 KPH	*−6*	*−9*	*−13*	*−17*	*−19*	*−23*	*−26*	*−30*	*−33*	*−35*	*−39*	*−43*	*−47*	*−50*
15 MPH	16	11	1	−6	−11	−18	−25	−33	−40	−45	−51	−60	−65	−70
24 KPH	*−9*	*−12*	*−17*	*−21*	*−24*	*−28*	*−32*	*−36*	*−40*	*−43*	*−46*	*−51*	*−54*	*−57*
20 MPH	12	3	−4	−9	−17	−24	−32	−40	−46	−52	−60	−68	−76	−81
32 KPH	*−11*	*−16*	*−20*	*−23*	*−27*	*−31*	*−36*	*−40*	*−43*	*−47*	*−51*	*−56*	*−60*	*−63*
25 MPH	7	0	−7	−15	−22	−29	−37	−45	−52	−58	−67	−75	−83	−89
40 KPH	*−14*	*−18*	*−22*	*−26*	*−30*	*−34*	*−38*	*−43*	*−47*	*−50*	*−55*	*−59*	*−64*	*−67*
30 MPH	5	−2	−11	−18	−26	−33	−41	−49	−56	−63	−70	−78	−87	−94
48 KPH	*−15*	*−19*	*−24*	*−28*	*−32*	*−36*	*−41*	*−45*	*−51*	*−53*	*−57*	*−61*	*−66*	*−70*
35 MPH	3	−4	−13	−20	−27	−35	−43	−52	−60	−67	−72	−83	−90	−98
56 KPH	*−16*	*−20*	*−25*	*−31*	*−33*	*−37*	*−42*	*−47*	*−51*	*−55*	*−58*	*−64*	*−68*	*−72*
40 MPH	1	−4	−15	−22	−29	−36	−45	−54	−62	−69	−76	−87	−94	−101
64 KPH	*−17*	*−20*	*−26*	*−30*	*−34*	*−38*	*−43*	*−48*	*−52*	*−56*	*−60*	*−66*	*−70*	*−74*

original wind-chill formula after studying the rate at
which bottles of water froze in Antarctica under different conditions of wind and temperature. Conditions
with equal freezing times were considered equivalent.
Wind-chill temperatures apply primarily to the rate at
which bare flesh freezes. Your bare skin has nothing but
its "boundary layer" of air to insulate it. A stiff breeze
can therefore remove 100 percent of your insulation
and create an effective wind-chill temperature that is
much lower than the air temperature. The risk of frostbite on a windy day is greatly increased compared to
that on a day with the same temperature and no wind.
If you're well-clothed, however, the "boundary layer"
of air might be only ten or 20 percent of your total insulation. Even a strong wind can only reduce your insulation by ten or 20 percent if you're wearing
windproof clothing.

Still shivering after your dip in the pond? Perhaps
you can build a fire. If you do, much of the heat you
feel will travel to you in the form of infrared radiation.
Hold a sheet of paper in front of your face, and you'll
notice an immediate drop in the heat you're receiving.
That's because the paper is blocking most of the radiation.

Radiation and conduction together account for nearly
all the heat passing through clothing. Radiation is actually a two-way process. Your warm body continually ex-
changes radiation with its cold surroundings. However,
the net flow of heat is always from warm to cold. In a
vacuum, radiation travels in a straight shot. In clothing,
however, radiation travels a tortuous path, emitted, absorbed and re-radiated over and over again as it
bounces about among the fibers. The denser the insulation, the more times the radiation must be absorbed

and re-radiated before escaping. Net heat flow is reduced. Using fibers of the right size also reduces heat
loss through radiation. The right size turns out to be
about ten microns. Some parts of a down plumule and
some microfibers have that diameter (more about this
later).

Another way to cut back the loss of radiant heat is
through placing a reflector in the clothing system. Compared to a clothing fiber, shiny metallic surfaces not
only reflect radiation, they also cut back on the radiant
heat emitted. Radiant barriers are most effective when
facing a dead air space. Practically speaking, that
means they must be either the outermost layer in a
clothing system or be integrated into a very low density
layer of insulation, such as a down or fiberfill jacket.

Furthermore, they must indeed be reflective, in other
words, shiny. While the "wet look" may once have
been fashionable at downhill ski areas, few people want
to look like a walking advertisement for Reynolds
Wrap. Any coating or fabric laminate to reduce the
space-oddity appearance or increase the durability also
decreases the value of the radiant barrier. Compression
by quilting the batting or squashing it beneath a pack
also reduces the efficiency, as does contact with any
conductive material. A wood stove provides a good
analogy. When made of blackened iron, the stove's radiation can be felt halfway across the room. If made of
shiny stainless steel, however, you would hardly know
the stove was hot—until you touched it. Conduction
would then immediately drain the stove's heat into your
hand. Coatings act like your hand, draining heat away.

Conventional fabrics, such as those used in long
johns and the various piles, are dense enough to deal
with the radiation component of heat loss pretty effec-

tively through absorption and re-radiation. Adding a reflective layer gives little additional warmth. Beware of advertising hype giving insulation values for some reflective barrier used alone. Adding the reflective barrier to a glove or pile jacket won't necessarily add the full insulation value of the reflective (measured alone) to the garment. The increase in insulation value will probably be much lower.

Reflectives do have some practical value in very low density bats of insulation, such as found in parkas and sleeping bags. Data from Mike Frankosky, a research scientist at Du Pont, shows that adding four reflective barriers to a Hollofil 808 polyester batting adds about 40 percent to the insulation value. If, instead of adding reflectives, you add a weight of batting equivalent to the weight of the reflectives, you only get 20 percent more insulation. Down on its own handles radiation better than the low-density synthetics, according to Frankosky, so you won't see as much benefit from reflectives. Radiation barriers are most effective if they're spread out. Grouped reflectives are less efficient on a weight basis.

Understanding modes of heat loss is most important in choosing a parka. Beneath your parka, however, you need one or several middleweight layers. For these layers, look for garments made of material that won't absorb water so they'll dry fast and retain their loft when damp. Wool's just too moisture-loving for my taste. So what does that leave you? Mostly garments made from that synthetic, furry-looking material called pile.

Pile fabrics are made using many techniques. Some give a fabric that has a furry appearance on one side, and a smooth, knitted surface on the other. Other techniques yield a fabric that is furry on both sides. The variations are endless and new methods are being invented all the time. Most piles are made of polyester, so that they're quite similar in their rapid drying and retention of loft when wet. The differences are in appearance, softness, resistance to pilling and wind resistance, though all require a windproof shell when Aeolus gets riled up.

Piles are also made of nylon and polypropylene. Generalizing about the virtues of the different materials is difficult because so much can be done during manufacture to alter the fiber's basic characteristics. Nylon in general is the strongest and most abrasion-resistant, with polyester second and polypropylene third. All three outlast wool. They also resist attack by insects, mildew and bacteria.

The trend in piles is to thinner layers, a plus for versatility and effective fine-tuning of your layering system. Wearing several layers also traps air in between the layers, leading to greater warmth than wearing one thick layer equal in weight to the total of the thin ones. However, one layer of the original thick stuff still goes into my pack for expeditions to truly frigid places like the Alaska Range.

If pile's so great, who needs a parka? The problem is that pile offers relatively little loft for its weight compared to the insulation normally used to fill parkas. Piles weigh several pounds per cubic foot, compared to half a pound per cubic foot for high-loft polyester battings, even less for high-quality down.

DOWN

On a warmth-for-weight basis, when dry, down is still the premier insulation. Its competitors, on a weight basis, are three kinds of polyester. PolarGuard is a continuous filament, which means the batting is made of many long, unbroken strands. The battings are sprayed with a resin to hold them together. Du Pont's Hollofil 808 is a staple fiber, which means it's cut into short lengths of two to three inches. Each fiber has a single hole running through its length. As Du Pont's Frankosky explains it, you build a high-loft batting by crimping each of the tiny fibers, then piling them on top of each other like a game of pick-up sticks with bent sticks. Putting a hole through each fiber, making it about 15 percent hollow, doesn't significantly reduce the fiber's stiffness and ability to make a lofty batting, but does give you 15 percent more polyester to make into fibers and pile on top. The result is that Hollofil battings are about 15 percent loftier for their weight than solid polyester battings. They're also more compressible—and more expensive. Hollofil II, a newer fiber, differs from Hollofil 808 in that it has a silicon-based surface treatment that makes it more compressible and resilient. It is also more expensive. Quallofil, also made by Du Pont, is another staple fiber which has four holes running through it. Du Pont estimates that Quallofil is five to eight percent more thermally efficient than Hollofil II on a weight basis.

None of the hollow fibers absorb water into their holes. In fact, according to Frankosky, it's almost impossible to *force* liquid in.

Assigning insulation values to materials is full of pitfalls related to the method of constructing the sample and the technique of measurement. However, researchers at Kansas State University tested all of these materials using a technique prescribed by the American Society for Testing and Materials. KSU found that good down is 2½ times as warm per unit weight as Polar-

Guard and Hollobond, a sandwich with resinated Hollofil 808 as the bread and Hollofil II as the cheese. KSU also found that good down is almost 70 percent warmer than Quallofil.

Other measurement techniques rank the insulators in the same order, but closer together. Du Pont's studies, for example, show that a down bag of weight equal to a Quallofil bag is 15 percent warmer. Note that Du Pont's results are closer to KSU's than they first appear because only half the weight of a typical parka or sleeping bag is fill weight. The greater thermal efficiency of down is diluted, so to speak, by the addition of the nylon shell, which weighs about the same whether it is containing down or Quallofil.

Most labs are quick to point out that insulation values measured on a flat hot plate (the technique KSU used) may bear little resemblance to the insulation of the finished garment. The garment's ability to hug your body as you move, referred to in the trade as its drape, also affects its warmth. A stiff garment allows pockets of air to form next to your body. When you move, that air is pumped out at the waist, neck and cuffs—the forced convection I described earlier. Garment design is important here, as is material and the way the fibers are processed into a batt. Most synthetic fibers, particularly those in resinated batts, do not drape as well as down. However, Quallofil's four-hole construction allowed Du Pont to crimp the fiber differently than they crimp Hollofil, giving Quallofil better drape and compressibility than Hollofil. A well-made Quallofil garment should have excellent drape. Pile drapes well, too, because its furry, protruding fibers fill air pockets. Handling a garment, then trying it on is probably the best way to assess a product's quality in this regard.

Down parkas can lose loft if you allow sweat vapors

to condense in them during many consecutive days of winter use and don't have the opportunity to dry them. Wearing a vapor-barrier shirt stops one source of moisture. Wearing a waterproof shell over the parka or buying a parka with a waterproof-breathable shell shields the parka from rain and melting snow. In some situations, however, it's still pretty tough to keep down dry indefinitely. Sailing in foul weather and sea-kayaking and river-running, particularly in northern climates, are prime examples. So is backpacking in cold, dank climates like coastal Alaska. In those situations, your best efforts to stay dry usually mean that you lose the battle against wetness more slowly than you would if you were careless. Synthetic parkas have the advantage that they lose less loft in perennially drizzly climates. However, the oft-repeated claim that "synthetics are warm when wet" needs qualification.

Professor W.C. Kaufman at the University of Wisconsin's Human Biology Department conducted a test in which he saturated a PolarGuard jacket and a down jacket by soaking and wringing them out repeatedly until they would hold no more water. Then he put the garments onto volunteers in a 40-degree room. Skin temperature readings were actually slightly lower for the PolarGuard than for the down. It's true that the PolarGuard did not collapse when wet, while the down did. The PolarGuard did indeed offer more resistance to heat loss by convection and conduction. However, the effect of evaporation completely overwhelmed any residual insulation provided by the PolarGuard. I would guess both garments felt quite chilly. The moral is that you have to keep your insulation from getting soaked no matter what it's made of. A synthetic jacket made of breathable nylon is only better than a down one if it's merely damp, not if it's completely soaked. However, if you can stop the evaporative heat loss somehow, such as by putting on a waterproof rain jacket, then the synthetic fiber's resiliency when damp would prove its worth.

The bottom line, for me, is that if it's going to be weather that only a mushroom person could love, I bring synthetics. For dry but cold conditions where weight is critical, down is still my choice.

Kaufman's data also shows that equal weight batts of down and PolarGuard, when saturated and wrung out, dried at the same rate. According to Kaufman, the synthetic batt held about three times its weight in water, the down about twice. The synthetic quickly dripped out enough water to reach the same weight as the down, which didn't drip at all. Both swatches then dried at the same rate because the same amount of water had to evaporate through the same kind of nylon shell.

Du Pont's tests, conducted under different circumstances, show that Quallofil dries three times faster than down. A close look at their chart of drying times, however, shows that they didn't wring out their samples. The actual drying *rate* is nearly the same in their test, but the total amount of water held is quite different. If they'd wrung out the water to the extent that Kaufman did, the drying times would have been much closer, although the Quallofil jacket would still have dried faster. Note also that it took three hours even for the Quallofil sample to drip-dry. It takes much less time than that to get thoroughly chilled, if not outright hypothermic. The motto, again, is to keep your insulation dry!

Compressibility is another factor to consider if you're

going to carry your parka on your back. According to Du Pont, Quallofil garments are just as compressible as down garments of equal warmth, and are even more compressible than down garments of equal weight. Hollofil is 20 to 25 percent less compressible, Du Pont says, and PolarGuard is less compressible still, by about 40 percent compared to Quallofil. Even the data from Reliance Products, the makers of PolarGuard, shows lower compressibility for PolarGuard. Under a load of 57.6 pounds per square foot, PolarGuard collapses to one third of its initial loft. Down collapses to one eighth of its initial loft. An unspecified "premium staple" collapses to about one sixth of its initial loft. As outdoor writer Mark Bryant once quoted a friend, "It takes strength to stuff a big PolarGuard bag. You need to have grown up milking cows."

While Du Pont's Quallofil statistics look impressive, in actual practice that loft and compressibility are hard to realize. I own both a Quallofil sleeping bag and a high-quality down bag stuffed with down that fills 625 cubic inches per ounce. My Quallofil bag has about three inches of loft, weighs 3¾ lbs and packs into a stuff sack that holds about 1,125 cubic inches. My down bag weighs four ounces less, has double the loft and fits into a stuff sack only ten percent larger. I wouldn't assert that the same is true of all Quallofil products because Quallofil can be put into a sleeping bag in several ways. Before you buy, check the weight (the manufacturer's catalog usually says), measure the loft, stuff the garment into its stuff sack and compare it to the other ones you're interested in.

Durability rankings among the synthetics are controversial. However, even Du Pont agrees that down outranks them all. According to their figures, synthetics will lose ten to 20 percent of their loft after several washings. Down, on the other hand, loses nothing if the down clumps are carefully broken up during drying. Down will also return to its original loft after compression for many more cycles. However, another factor to consider is the durability and care of the shell fabric. Many parkas with either man-made or down insulation are retired long before the insulation itself is worn out, simply because the shell fabric will no longer hold the insulation together. Synthetic parkas are much cheaper initially than down ones, but well-cared-for down is usually no more expensive in the long run and may actually be cheaper.

There's a whole other class of "thin" insulators available which give about 60 to 80 percent more insulation for a given *thickness* than high-loft polyester battings. The insulation values are usually given in clo per centimeter of insulation thickness. One clo is roughly equivalent to the insulation provided by a man's business suit. Looked at another way, if you're comfortable when sitting, unclothed, in a room at 86 degrees, you would also be comfortable wearing one clo of clothing while sitting in a room at 73 degrees.

Thin insulators give about 1.6 to 1.8 clo per centimeter, depending on who's doing the measuring. Hollofil gives about 1.0 clo per centimeter. Thin insulations are made of microfibers only two to ten microns thick— much thinner than the 25-micron fibers used in high-loft battings. These insulations work so much better than high-loft battings on a thickness basis because their higher-than-average density serves as a better-than-average barrier to the loss of radiant heat. In addition, their fibers are the right size to effectively scatter the wavelengths of infrared radiation most commonly produced by the human body. Radiation must take a very circuitous path through a thin insulator before escaping.

The drawback of thin-fiber insulators is their greater density—their greater weight per unit of warmth. Figures from Du Pont comparing two batts of insulation of equal weight, one made of Thermolite (a thin fiber insulator), the other of Quallofil, show that Thermolite gives only 60 percent of the insulation. Thin-fiber insulations are useful where bulk is a drawback and weight is not critical. Gloves are one example; fashionable ski clothes are another. Some mountaineers in extremely cold environments, for example during the first winter ascent of McKinley's Cassin Ridge and an ascent of the Himalayan peak Cholatse, felt the bulk of the high-loft garments they would be forced to carry was so great that they preferred some extra weight to reduce the mummified feeling. In general, however, high-loft polyester insulations or down are more useful in the backcountry where weight is the prime concern.

Regardless of the filling you choose, a good parka should reach below your buttocks to conserve the large amounts of heat produced by the muscles there. Parkas with elastic at the bottom hem tend to ride up to belt level, reducing their value. Most down parkas, even the good ones, have sewn-through seams: places where the inner and outer shells are sewn directly to each other, with no insulation in between. That's fine if the compartments are well filled with down. To compare the filling of different parkas, hold them up to a strong light.

Good hoods fasten easily, even with gloves on. Some are designed to cover your mouth; others fasten below your chin. I much prefer those that fasten below my chin so my breath doesn't condense in the insulation.

Thoughtfully designed parkas also have differentially cut shoulder areas. The outer layer in the shoulders is cut larger than the inner layer, so the weight of the jacket is supported by the inner shell. This gives the insulation in the shoulder area room to expand. In jackets without this kind of cut, the shoulder area insulation is compressed by the jacket's weight.

My favorite sleeve design features knit cuffs recessed inside the end of the sleeve for protection from snow. Unprotected knit cuffs collect snow quickly. Elastic cuffs don't seal warm air in as effectively. Parkas with waterproof-breathable outer shell fabrics stay drier longer.

The best parka in the world won't keep you warm if it's saturated by rain or melting snow. The best pile suit won't keep you warm if the wind is driving its icy fingers through every square inch. To stop the wind, rain and snow, you need a good shell jacket and pants. That's what I'll consider next.

· 4 ·
THE SHELL GAME:
Clothing to Stop Wind, Rain and Snow

Half-blinded by the driving sleet and rain, I thought for a moment I was watching Napoleon's retreat from Moscow. Hikers in an endless line snaked down the trail toward timberline 2,400 feet below. A few were well equipped, with snug rain jackets and waterproof pants. Far too many, however, were flatlanders unprepared for the vicious summer thunderstorms that plague Colorado's 14,256-foot Longs Peak. The frigid rain had already soaked most of them down to their cotton underwear. For them, a simple sprained ankle would have turned discomfort into serious trouble. Although the calendar said August, the temperature was in the 40s. With the wind chill, it was well below freezing.

Clothing soaked by rain or melting snow will chill you faster than anything besides a dunking in a mountain lake. When the wind gets rowdy, even dry clothing should usually be covered by a windproof shell. Shells add significant warmth even in still air. Tests by REI showed that adding an uninsulated shell parka to a pile jacket allowed the wearer to remain comfortable when the temperature in the test chamber was dropped eight degrees. *Anything* to keep off that heat-robbing wind, rain and snow can stave off the shivers or worse. In a pinch, I've even cut head and arm holes in a plastic garbage bag and worn it as a vest.

As any experienced outdoorsperson knows, staying warm and dry in a cold, steady rain demands good gear and a lot of savvy. In fact, it's often more difficult than staying warm in a snowstorm simply because rain will eventually penetrate the tiniest weakness in your shell gear. Snow won't.

Before buying a shell garment, decide what conditions you're likely to encounter. In the cold, dry climate of the wintertime Rockies, I got away for several years with simple, inexpensive, uncoated nylon wind pants and jacket. They blocked wind pretty well and also "breathed," in other words, let some sweat vapor escape. True, if I sat in the snow for very long, I got wet. In addition, the fabric wasn't as windproof as the new fabrics that are both waterproof and breathable. High winds reduce the insulation provided by any clothing system by increasing convection at the outside surface and by forcing cold air through the outer shell. However, a very windproof fabric reduces the chilling effect of wind. If your shell is totally windproof, you're left with 20 to 40 percent more insulation in high winds than you'd have with a conventional nylon shell, according to W.L. Gore, makers of the first popular waterproof-breathable, Gore-tex.

Based on my experience, I'd have to agree with Gore's claims. When I first started using a Gore-tex jacket, I was convinced it didn't breath simply because I got so hot in it while wearing the same number of layers I'd typically worn under my old nylon shell gear. Then I realized its windproofness was the cause. Breathability is roughly the same, according to Gore. Removing one insulating layer stopped the overheating.

If all you need is water- and wind-resistance, consider one of the new "water-resistant-by-construction" fabrics. These fabrics are woven extremely tightly from very fine yarns, most often polyester, but sometimes a polyester/nylon blend. The breathability for most is high, in the range of the waterproof-breathables and uncoated nylon. I'll get into breathability statistics when I talk about waterproof-breathables.

These new, extremely tight-weave fabrics resist water better than ordinary nylon, but that doesn't make them good rainwear. They'll keep off mist and drizzle, but not a tropical monsoon.

Water resistant, by the way, is a stronger term than water-repellent, which is usually just a treatment to make water bead up on the surface of a fabric. The rule of thumb used to distinguish water-resistant from waterproof fabrics at the Army's Research Institute of Environmental Medicine in Natick, Massachusetts, is that a waterproof fabric should not leak when water exerts a force of 25 pounds per square inch. The Mullens Hydrostatic Test is the most common waterproofness test today.

If 25 psi seems like overkill, consider the pressure exerted on fabric when kneeling on wet ground. Consider, too, the pressure exerted by pack straps as a heavy load shifts back and forth from shoulder blade to shoulder blade. In addition, the initial performance of the fabric must be good enough so that no leakage occurs if the fabric is stretched when water pressure is applied. For its heavy-duty rain parka and trousers, Natick specifies that the fabric withstand 250 psi water pressure. Fabrics that waterproof can rival medieval armor in stiffness and aren't much more comfortable to wear.

Water-resistant-by-construction fabrics typically withstand one psi or less. Manufacturers usually give the number in a different form, however: in millimeters, centimeters or inches on the Suter Test. The number refers to the height of a column of water that the fabric can withstand without leaking. Converting Suter test results to psi is easy. Begin by converting the Suter-test number to inches, if necessary. If the number is in centimeters, multiply by .4. If it's in millimeters, multiply by .04. Then multiply the height in inches by water's density, .036 pounds per cubic inch. That figure is the pounds per square inch the fabric will withstand.

(For those of you who are mathematically inclined, the formula is:

$$\text{Psi} = \frac{\text{density of water} \times \text{height of column} \times \text{area of column}}{\text{area of column}}$$

The area of the column cancels out, meaning the waterproofness in psi is simply the density of water in pounds per cubic inch times the height of the column in inches.

If you're seriously concerned about rain, or want something that will let you kneel or sit down in the snow, you need gear that resists more than one psi. Raincoats made of sheets of vinyl with heat-sealed

seams and no strength-giving fabric fill the cheapest end
of the waterproof spectrum. They serve well for fetch-
ing the mail from your mailbox, but don't have the durability necessary for anything more adventurous.

Urethane, polyvinyl chloride, chloroprene, chlorosulfonated polyethylene, and nitrile rubber are commonly used to waterproof rainwear fabrics. Chloroprene is better known by its Du Pont tradename,
Neoprene. Chlorosulfonated polyethylene is also better
known by its Du Pont tradename, Hypalon. In most
cases, the fabric is nylon. Coating names should be regarded as general categories rather than specific products because so much can be done to modify each basic
type. It's tough to generalize, but a few broad state-
ments hold true.

None of these coatings, as normally used, permit the
fabric to breathe. All are used in sailor's foul-weather
gear. Sailors, however, impose different demands on
their gear than do mountaineers, skiers and backpackers. Far more water deluges sailors in rough
weather than descends upon a hiker in a gentle rain.
Sailing also grinds up foul-weather gear a lot faster than
backpacking. In addition, sailors don't need to carry
their foul-weather gear on their backs except when
they're actually wearing it. Sailors lean toward very
heavy but very durable PVC, Neoprene and nitrile rubber suits. By contrast, backpackers looking for purely
waterproof gear most frequently select the synthetic
rubber called urethane.

Urethane's major advantage for hikers is that a waterproof seal can be obtained with a thinner coating
than is required with Neoprene, PVC or nitrile rubber.
Less material applied per yard of fabric generally means
lower cost, lighter weight and greater flexibility. Du-

rability is variable because urethane can be formulated in many different ways. In general, softer, more drapable coatings are also less durable. Stiffer, harder coatings last longer. Urethane can also be formed into a sheet, then laminated to different backing fabrics. Don't store a wet, urethane-coated garment in a hot environment, such as a car trunk. The urethane can peel off the base fabric or start growing mildew, a fungus which needs heat, moisture and dirt to multiply. The mildew (or its by-products) can make the coating leak.

Condensation is a potential problem inside any rain garment, whether it's billed as waterproof-breathable or not. Raingear adds a lot of warmth to your clothing system. In many situations, the best solution is to replace one of your warmth layers with your raingear, rather than adding the raingear on top. If you can't bring yourself to do that, then pause to remove clothing at the first sign of overheating. Using a vapor-barrier shirt alerts you to overheating at the first possible moment and prevents moisture from ever reaching your shell jacket. I wore a totally waterproof shell jacket over a vapor-barrier shirt during a first ascent in April, 1983, on the north face of Alaska's 17,400-foot Mt. Foraker and stayed warm and dry. It's simply not true that coated fabrics make you sweat, or that breathable fabrics keep you dry. Walk outside on a 20-degree morning wearing nothing on your torso but a vapor-barrier shirt and see if you sweat. Walk outside at noon in July in your breathable down parka and see if your skin stays dry. Solving the sweat problem relies more on your regulation of body temperature than on the fabric you're wearing.

But, let's face it, it's pretty hard not to sweat a little

once in a while. Recognizing this, manufacturers have continually tried to invent fabrics that are both water-proof *and* breathable.

All the waterproof-breathables currently available use either the microporosity principle or the absorbent-film principle. Both types of fabric work because of the difference between water as vapor and water as liquid.

Water droplets are composed of molecules held to-gether by an attraction called the van der Waals force. Water vapor, on the other hand, is composed of single water molecules energetic enough to overcome the van der Waals force and bounce away from each other, forming a gas. Microporous waterproof/breathable fab-rics have pores big enough to permit single water mole-cules in vapor form to pass through. A water molecule is roughly .0002 microns across. The pores range from .2 to 2 or more microns, depending on the fabric. The pores are small enough, however, to prevent the pas-sage of liquid water droplets so long as the microporous film or coating is hydrophobic—in other words, so long as drops of water tend to bead up on the microporous surface rather than wick through the holes. Some mi-croporous fabrics leak when contaminated by oils from your skin, insect repellent, soap residues or sunscreen. These agents change the surface characteristics of the film or coating so that water tends to wick (just as it does along treated polypropylene and polyester) rather than bead up. The water eventually wicks through the holes, soaking the wearer.

Gore-tex, the first waterproof-breathable, originally worked solely on a microporosity principle. It was made of a sheet of polytetrafluoroethylene, better known by Du Pont's trademark, Teflon. The sheet was stretched to make it porous, then either laminated to one protective fabric, which became the jacket's outermost layer, or glued to two fabrics with the Gore-tex in the middle.

Many of its competitors, such as Entrant, are made of fabric coated with a layer of microporous polyurethane. The manufacturers create the pores by adding a foaming agent when the urethane is still liquid. So far, all the foamed urethane coatings are less waterproof than Gore-tex.

Absorbent-film waterproof-breathables have no pores. Instead, the extremely water-loving film, which is .001 of an inch thick or less, absorbs water molecules on the hot, sweaty side, passes them through the film by diffusion and releases them on the cooler, drier side. In theory, no liquid water penetrates, even when the fabric gets contaminated, because no pores exist as little tunnels. However, these fabrics rely, at least in part, on the water-repellent finish on the outer fabric to keep liquid water away from the film. A sheet of water on the outside of the film might, as a minimum, reduce the film's breathability. In some circumstances (such as sitting in a puddle of warm water on a boat) water vapor might actually flow from the outside in. Microporous fabrics also rely, in part, on the water-repellent finishes on the outer fabrics to prevent a reverse flow of water vapor.

Bion-II was the first waterproof-breathable to advertise that it worked on the absorbent-film principle. However, a look at W.L. Gore's patents, plus a conversation with Sam Brinton at Gore, reveal that all the Gore-tex produced since 1978 is actually a bicomponent membrane. One part is still the stretched Teflon. The other part is a non-porous film coating that works just like Bion-II.

In theory, waterproof-breathable fabrics should create a hiker's nirvana. In practice, there are limitations. As one industry wag put it, "Gore-tex is not an antiperspirant." He could have been referring to the other waterproof-breathables as well. To understand the limitations of these fabrics, we need to make a fast foray into physics.

The first limitation involves the force required to drive sweat in the form of water vapor through the fabric. Waterproof-breathable fabrics can't "pump out" water vapor. Nor do they work by permitting air to flow easily through the fabric, taking sweat vapor with it. In fact, all of these fabrics boast, rightfully, of their windproofness. Instead, water vapor moves through the fabric only when driven by a vapor pressure gradient.

Picture a closed box filled with water vapor but no air. The water molecules are bouncing off the box's walls and exerting a pressure, called the vapor pressure. Raise the temperature and the molecules get more excited. They bounce off the walls harder and more frequently, and the vapor pressure goes up. Cool the box, and the opposite happens: the vapor pressure drops.

Now add air to your imaginary box. *The vapor pressure exerted by the water molecules remains the same.* Warm the box, and the water vapor pressure goes up; cool it and the vapor pressure drops.

Now let's say you've got two boxes butted up against each other. Only a waterproof-breathable membrane separates them. Water molecules are banging into and passing through the membrane from both sides, but the box with the greater vapor pressure (the warmer one) has more energetic water molecules, so more of them bang into the membrane and pass through it each second. A net flow of water molecules begins. The difference in vapor pressure between the two sides is the vapor-pressure gradient.

Waterproof-breathable fabrics "breathe" only when the vapor pressure is greater inside the fabric than out. The greater the vapor pressure gradient, the more vapor passes through the membrane per hour. The vapor pressure gradient is greatest when the fabric is close to the heat source, your 90-degree F skin, and when the air outside is cool and dry. If you're sweating inside the garment in a steamy Georgia thunderstorm when the temperature inside the garment and out is 90 degrees, and the relative humidity inside and out is 100 percent, no sweat vapor will escape at all.

The temperature difference between the space just inside a shell jacket and the outside is pretty constant, regardless of the outside temperature, if the wearer's activity level is constant. In cold weather you need more clothing underneath the jacket than you do when it's warm, but that doesn't affect the temperature drop from the inside to the outside of the shell itself, disregarding the other insulation. Unfortunately, the vapor pressure gradient produced by that near-constant drop in temperature is *not* constant. If it's 42 degrees F just inside the jacket and 32 degrees F outside, you've got less than half the vapor pressure gradient that you'd have if it was 80 degrees just inside the jacket and 70 degrees outside. That's assuming the relative humidity inside and out is 100 percent. If the outside humidity is lower, which it usually would be, the difference in vapor pressure gradients is even greater. In cold weather you've got only one-fourth to one-half the "breathability" that you have when the temperature is mild. Since it's easy, even in warm weather, to produce more sweat than a waterproof-breathable fabric will pass, you can imagine how sweaty you can get in the cold if you don't regulate your temperature by removing insulation layers.

A further problem in cold environments is condensation within the insulation between your skin and the shell garment. If the temperature drop between skin and environment is steep enough, most of the sweat vapor will condense long before reaching the shell. To take just one example, if the temperature just inside your uninsulated shell jacket is 32 degrees F, as it could easily be when the outside temperature is ten or 20 degrees colder, then roughly 85 percent of the moisture produced by your body will have condensed by the time it reaches the shell. Liquid water, of course, won't pass through any waterproof-breathable. In real cold, the most breathable fabric in the world does little good. A vapor-barrier shirt is required to prevent your insulation, particularly if it's down, from losing its effectiveness. To gain some idea yourself of waterproof-breathable fabric's limitations, pour boiling water into a jar. Cap the jar with a piece of fabric held on with a strong rubber band. Place the jar in the freezer. Within half an hour the inside of the fabric will be a sheet of ice.

Manufacturers have recognized that condensation can occur even inside a waterproof-breathable fabric. They have begun selling various kinds of lining fabrics intended to soak up the perspiration and spread it out through wicking so, at least in theory, it can escape more readily. Wicking in underwear is touted as a virtue for the same reason. The fanciest of these new liners are bicomponent, with a hydrophobic ("water-hating") layer inside, towards your skin, and a hydrophilic ("water-loving") layer outside. In theory the inner layer always feels dry against your skin (though you could only tell if you were wearing a t-shirt or less), while the outer layer wicks very fast. Although these liners may have some virtues, particularly in mild weather, I try to avoid sweating hard enough to give them a chance to do their job.

How breathable is breathable enough to do any good? It's almost impossible to compare breathability statistics. Although most companies give the results of their moisture vapor transport rate (mvtr) tests in grams per square meter per 24 hours, the tests producing those results come in widely different forms. One company's 14,900 gm/m^2/24 hrs may be the same as another's 1,200 gm/m^2/24 hrs. Usually the method is not specified. Even if it is, it's pretty hard to translate the numbers into comfort in the field.

The bottom line, however, is that none of these fabrics will keep you from overheating and sweating if you're wearing too much clothing. If you do let yourself sweat, wearing waterproof/breathable gear will probably let you dry out a little faster, particularly under mild conditions, than if you're wearing waterproof gear. If you're careful about heat regulation and ventilation, you can probably be as comfortable in purely waterproof gear as in waterproof-breathable gear, particularly in the cold.

DURABILITY

No fabric remains waterproof forever. Coatings abrade and eventually leak. The glue dots holding the Gore-tex membrane to its fabrics put stresses on the membrane when the fabric is flexed during use or when stuffed into a pack. Eventually, the membrane can get small tears and start to leak.

In general, the heavier the coating, the greater the durability. Some coated rainwear has a very light scrim laminated to the coating on the inside. This scrim should help reduce abrasion as well. Likewise, the heavier the fabric protecting the Gore-tex membrane, the more durable it will be. The three-layer Gore-tex laminates where all layers are glued to each other tend to be more durable than the two-layer versions with free-floating liner, or the three-layer versions where the Gore-tex is laminated to only a free-floating central layer. Don't take a running suit made of ripstop weighing an ounce a square yard to Everest, then expect to stay dry when you hike out through the monsoon afterwards.

PONCHOS, CAGOULES, AND RAINSUITS

No material, no matter how waterproof, will keep you dry if it's sewn into a poorly designed and constructed garment.

For starters, forget about ponchos for winter use or serious rain. Ponchos are large, rectangular sheets of waterproof material with a hole in the center for your head. Most have hoods and are large enough to fit over a small pack. They offer good ventilation, low cost and very few seams, which are always potential leak points. If the wind kicks up, however, you'd better sprint for your car. Wind lifts the bottom of the poncho, obscuring vision of your feet and letting wind-driven rain and snow soak you from the waist down. In addition, tree limbs have the same affinity for the flapping fabric that they do for Charlie Brown's kite.

Cagoules give somewhat better protection, but not by much. Cagoules are mid-calf or ankle-length, hooded, pullover garments with a short zipper at the neck.

Though they have less tendency to lift in a wind, they still obscure your feet and often shorten your stride. They will protect your legs longer than a poncho, but not as long as a pair of rain pants. The extra length can be pulled over your feet during an emergency bivouac, but carrying a cagoule for that supposed advantage is like carrying a sleeping bag on every two-mile day hike "just in case."

Far and away the best shell gear for cold, snowy or rainy weather is a jacket and pants. I learned the hard way that pants are just as important as a jacket. While scrambling up Longs Peak one August day I made the mistake of ignoring the growing thunderhead rumbling directly toward me. I was at 14,000 feet when the storm struck. I donned my jacket and headed down as the fearsome updrafts roaring up the steep slopes swept freezing rain, hail and snow into my face. The deluge immediately soaked my cotton-polyester pants. I'd put them on in the morning thinking, this is summer, right? No need for the full winter regalia. As I slid on my butt down ice-coated rocks, then started running for timberline, I swore I'd never venture into the high mountains again in any month without full protection.

Shell garments come either as pullovers with short neck zippers or as jackets with full-length front zippers. Both are often called anoraks from the Eskimo word meaning a jacket with a hood. The pullover style resists wind-driven rain longer because of the shorter zipper. The full-length zipper in the jacket gives better ventilation and easier access to inside pockets. It's generally preferable for winter use because snow won't penetrate a well-protected zipper.

An alternative to full pants is rain chaps, which are constructed like two separate pant legs. They provide the cheapest rain protection for your legs, but don't cut it in the winter because you lose too much heat and give snow too many opportunities to creep inside. If you buy chaps for rain, make sure your rain jacket hangs low enough to shelter the uncovered areas in the crotch.

Whether you buy chaps or full rain pants, make sure they come down over the tops of your boots. If they stop short, water runs off your legs and turns your boots into bathtubs. If you end up with short pant legs for some reason, you can beat the runoff problem by wearing waterproof gaiters with your pant legs outside.

True rain pants usually have a short ankle zipper that is often backed by a triangular gusset of fabric. They usually won't fit over double boots. Shell pants for winter use or expeditions, on the other hand, should have full-length separating side zippers so that the pants can be put on over double boots, skis and crampons and be ventilated from the waist down. Using expedition pants in heavy rain will probably lead to leakage through the zippers.

The need for seams gives water one of its biggest advantages. Try to avoid jackets with seams directly across the top of the shoulder, where rain beats down directly. Factory-sealed seams, in which a strip of seam tape is permanently glued on by heat and pressure, are best, although factory sealing is not always perfect. All hand-applied seam dopes require careful application and frequent renewal if you want to stay dry. Apply them when the fabric is stretched to open up the pinholes where the needle went through. Heavyweight PVC foul-weather gear often has electronically welded seams, which are probably the most water-tight seams available.

The ideal hood may be lurking out there somewhere,

but I haven't found it yet. Most give you sterling views of your hood lining whenever you turn your head. If you need goggles because of blowing snow, try wearing them over the hood to hold it to your head when you move. Unfortunately, holding your hood in place by wearing goggles over your forehead instead of your eyes usually causes the goggles to ice up from sweat condensation. When trying on different jackets, look for a hood that accommodates your head size and neck length.

Hoods that keep your forehead dry with a visor also restrict your vision upward. Hoods that grant you better upward visibility give proportionally poorer protection. Allowing room for a ski hat underneath your hood often means that the hood used without one almost covers your eyes when cinched down with the drawstring. I usually do allow room, however, because most of the time that I'm wearing my hood up I also want a hat.

For best rain protection, two storm flaps should cover the front zipper. They may be outside, with one facing left, the other right, or one may be inside, the other out. One flap is fine in snow country. Flaps that are held down with hook-and-loop are usually more watertight than those held with snaps. They are also much easier to manipulate with cold, wet fingers or gloves. Avoid storm flaps that constantly bind in the zipper.

Pockets are another potential crack in the dike. In raingear, the fewer pockets the better. Cargo pockets, usually found low on the jacket on either side of the zipper, are close to worthless. During storms, when not wearing a pack, they fill with water or snow. While wearing a pack at any time the waist belt completely blocks access to the pockets and jams the contents into your hips. My preference is for zippered pockets situated high on the jacket's front panel. They should be inside the pack shoulder straps, beneath the pack's sternum strap and above the waist belt.

Cuffs should be cut wide for ventilation with the option of cinching the cuff closed with hook-and-loop so you can reach over your head and not have your sleeve turn into a downspout funneling water into your garment. Constantly closed knit or elastic cuffs inhibit ventilation too much.

Roomy rain jackets are easier to ventilate than tight ones because they pump out larger amounts of moist air when you move. In addition, they encourage a stronger "chimney effect," in which cool outside air enters at the waist, flows upward and escapes at the neck, taking moist air with it. That can help reduce condensation. Wearing a pack with a waist belt stops any chimney effect, of course, as well as closing down chest and back vents. In that situation you can only ventilate by opening the front zipper and the armpit zippers, if you've got them. Both actions let in snow and rain if it's really dumping. The best cooling system when you're wearing a pack is usually just removing an insulating layer.

When trying on shell jackets, put on a loaded pack and lift your arms. Some shoulder/sleeve designs give you more freedom of movement than others. Try touching your elbows together in front of you to check for freedom in that movement. Can you reach forward without the sleeves pulling up and exposing clothing beneath?

You actually already own a complete suit of the best shell gear ever created—human skin. Its waterproofness and vapor permeability are unquestionably topnotch. Skin also has another advantage: when damaged, it repairs itself. Regrettably, nothing man-made can, as yet, quite match it.

· 5 ·
ONLY TWENTY DOLLARS PER TOE:
The Secrets of Warm Feet

After a week of 20-below weather, Tom Reid figured Upper Titcomb Lake, in Wyoming's Wind River Range, would be frozen solid. He figured wrong. Flowing water near the lake's outlet had kept the ice thin. Without warning, the ice shattered beneath his skis.

He threw himself backwards and landed on solid ice, soaked to his knees. Instantly his clothes turned to ice. He sprinted for the snow cave some skiers had built four miles away. Luckily, he arrived unscathed.

That night he dried his wet socks and low-cut touring boots in his sleeping bag. By morning only the boots felt damp. He decided to ski to his car, 18 miles away.

After five miles of skiing, with the temperature at 25 below, the molded plastic soles of his boots cracked and began to fall apart. Snow crept inside, melted, then froze. Still, he thought he could reach his car without frostbite. He hurried on.

Just before sunset his toes went numb. At 8 p.m. he reached his car. Six toes felt like marble. Within hours they blistered black. Not even prompt medical attention could repair the damage. Two months later he honed the tip of his friend's sharpest buck knife and severed the last scrap of dead flesh holding on what had once been the two smallest toes of his right foot.

Good footwear may seem expensive at first glance. High-quality double boots for skiing and mountaineering can cost $200 and up. Still, as veteran boot repairman Steve Komito puts it, "Good boots are actually cheap. They cost only $20 per toe."

Don't pay cheap footwear's true price. Good boots pay for themselves in the long run. The first step in selecting cold-weather footwear is to figure out what terrain you'll be encountering, the temperatures you're likely to meet and how you plan to get around. Hikers need a boot that flexes at the ball of the foot; mountaineers want one that doesn't. Skiers need a boot that fits their bindings; snowshoers can get away with almost anything.

You also need to learn, from experience, how much insulation you need. Some people are apparently born with sluggish circulation in their toes. Others have abused their feet in the past by letting them go numb repeatedly, which slowly but surely damages the foot's circulatory system. My friend Janet spent her teenage years ski-racing in tight boots she'd cranked down to the limit for better control. Her feet had been constantly numb. The bad habit bit back when we went to McKinley in April in identical boots with identical socks. My feet (I grew up surfing in Southern California) were fine; hers were always painfully cold.

Lightweight fabric-and-leather hiking boots fill the gap between running shoes and more substantial foot-

wear. Most use a cemented-sole construction, in which
the bottom edge of the upper is folded inward beneath
the insole and glued. The outer sole or tread is glued on
last. These boots may be fine for moderate cold if you
buy them big enough to fit with two pairs of socks.
However, water easily penetrates the myriad seams and
you can't use ordinary boot waterproofings on the fabric. When the seams abrade, the boot usually falls apart
fast even if the fabric and leather pieces themselves are
still intact. If you don't mind wearing zebra shoes, you
can lengthen the boots' life by coating the seams with a
boot-patching compound.

W.L. Gore makes a Gore-tex boot liner for fab-
ric/leather and lightweight all-leather boots out of a
membrane that's supposed to be twice as thick and four
times as tough as the apparel membrane. Boots take a
lot of punishment, however, so don't expect a membrane only .002 inches thick to remain waterproof forever.

More durable in appearance, but actually worse, are
boots with leather uppers and injection-molded soles,
like the ones that cost Tom Reid two toes. Boot-makers
fabricate these by suspending the upper above a mold
shaped like the finished sole. Then they inject a liquid,
often polyvinyl chloride, into the mold to create a one-
piece sole with the lower edge of the upper trapped inside. When any kind of injection-molded boot fails, it
usually fails catastrophically—the sole either cracks or
separates from the upper. Repair and resoling are almost impossible because glue bonds poorly to the PVC.
A few manufacturers still make ski-touring boots this
way, but most lightweight ski-touring boots and all
lightweight hiking boots now are either cemented or cemented and stitched.

Lightweight boots are now so flexible, and gluing has

gotten so good, that boots held together only with glue are reasonably durable. However, a stiff leather boot that has to withstand really hard use, particularly a ski-touring boot, should still be held together with stitching as well as glue.

Stitched boots can be either inside fastened or out-side fastened. To make an inside-fastened boot, the bottom edge of the upper is turned inward and stitched to the insole and midsole. The technique, also called McKay or Littleway construction, gives good durability in medium-weight footwear. In one hybrid form, the upper is stitched to a molded plastic midsole, then glued to the outer sole. In a ski-touring boot, this means that the portion of the boot clamped into a three-pin ski binding is attached only to the rest of the boot by glue. That glue joint can fail if you crank your turns really hard.

Boots built of leather too thick for inside-stitching machines to handle have a Norwegian welt. One row of stitching runs horizontally, connecting the bottom edge of the upper, which is turned outward, to the edge of the insole, which is turned downward. A second row of stitching runs vertically through the edge of the upper and the midsoles. Almost no hiking boots use this tech-nique nowadays, but you'll still see it in the best ski-touring boots.

Durability is only related to warm feet in that boot failure far from civilization can be disastrous. Heavy boots aren't necessarily much warmer than lighter ones because the leather, while thicker, isn't thick enough to give much added insulation. Adding a pound to your feet costs as much energy, when you're moving, as adding 4½ to ten pounds to your back, according to the Army's Natick research lab. Heavy boots may be drier

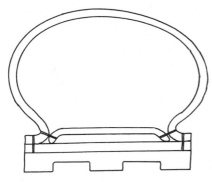

Cross-section of outside-fastened or Norwegian-welt boot.

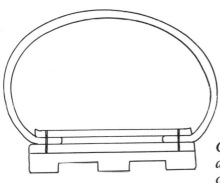

Cross-section of inside fastened boot, also called McKay or Littleway construction.

than lighter ones, however, if the reason the boot is heavier is a difference in leather.

Split leather is a layer of hide which doesn't contain the skin surface, called the grain. Typically splits are weaker and less waterproof than full-grain leathers containing the skin surface. However, a good split will outperform a poor full-grain. Good leather of either type costs a lot, so price is a pretty good indication of quality.

Heavy boots use full-grain leather for its durability and water resistance. Here debate rages over which way the grain side should face. Cobblers agree that the most waterproof layer of the hide is the grain because it's the most dense. Some think boots should be made rough-side out so the grain surface is inside and protected from abrasion. Others think rough-side-out construction allows the less dense inner surface of leather to absorb water. A boot built well of good materials in either style will keep your feet dry for a long time.

But not forever. No leather boot will keep your feet dry if you're standing in water all day. For that, you need a rubber boot.

DOUBLE BOOTS

If your feet are dry but cold, you need a boot with more insulation. You can't keep adding socks inside the same boot, however, because you quickly begin to constrict circulation. That's worse than having fewer socks. George Leigh-Mallory reputedly once stopped high on Everest to remove a pair of socks so his feet could warm up. You also can't just buy a bigger boot and add more socks, beyond a certain point, because the socks compress too much. Your foot walks around inside the boot instead of outside, and you lose all control. You need more insulation, but in a form that won't compress readily. You need a double boot.

Double boots are just that: a soft, insulated inner boot with minimal tread or none, and a rugged, usually uninsulated, outer shell. The inner boot is usually made of wool felt or foam and lightweight leather (or a synthetic substitute). The very lightest inner boots today are pure Alveolite foam. Though warmer than felt inner boots, this foam also packs down with time. The inner boots belonging to my partner on McKinley's Reality Ridge compressed so much that he could hardly climb. He had to stuff another extremely thick pair of socks inside to regain control.

You'll only find double boots with outer shells made of leather in the backcountry skiing and telemarking department, where flexibility is still required. All leather double boots cost plenty, will keep your toes toasty in all but the most extreme conditions and will last for years with proper care.

Almost all mountaineering double boots, on the other hand, have plastic shells. Mountaineers want rigidity for ice-climbing. Plastic provides it. These plastic double boots weigh several pounds less than the leather double boots they replaced. They also won't freeze into steel shells, like leather boots will if you let them get wet either from the outside or from sweat not blocked by a vapor-barrier sock. Although plastic double boots seem incredibly tough, a few of the outer shells have cracked. You can't repair a cracked shell in the field (or in the repair shop, for that matter). All you can do is tape up the gap and head for home.

BOOT CARE

Leather boots—if you take care of them right—won't do that. Heat and dirt are a leather boot's worst enemies. Dirt works its way into the stitching. The boot's flexing as you walk or ski then grinds up the thread. If your boots get muddy, clean them with a scrub brush and warm water and dry them at room temperature. Never get boots hotter than your own skin can tolerate. Heat damages leather. Heat is also what boot repair people use to get your old soles off before resoling. To avoid a premature resoling bill, don't leave your boots in the sun in a hot car.

Leaky boots chill your feet as much as leaky raingear chills the rest of you. The best boot waterproofings are made of beeswax, lanolin, fish oils and solvents to help the waterproofings penetrate. They keep water out by filling up leather's pores. They also condition the leather, keeping it supple. Don't lather on any of them. Your boots will turn into dishrags. In addition, the extra gunk will collect dirt, particularly along the welt, which will wear on the stitching. Neat's foot oil softens leather excessively and should be avoided. Silicone waterproofings, while effective, last only briefly.

SHOE-PACK AND VAPOR-BARRIER BOOTS

So far we've been talking about double boots intended to keep you from spraining your ankle while packing a load or give you control over your crampons or skis. If you don't need that kind of support, consider shoe-pack boots. These usually have a molded rubber foot section and a high-topped leather upper. Inside lies a thick, removable felt liner. They're totally waterproof up to the stitching, pretty waterproof above that, and very warm. I'd hate to tackle anything more rugged than a staircase in them, however.

Vapor-barrier boots, also called moon boots, Mickey Mouse boots, Korean boots and K-boots, are heavy, bulbous rubber monstrosities that offer extremely good insulation and no maneuverability. They're the sort of boot you'd wear surveying in Antarctica. If the airtight outer shell is breached, the boots become worthless. You can find these boots occasionally in Army surplus stores.

The best boot around isn't worth its weight in shoelaces if it doesn't fit. For cold-weather use, make sure you've got enough room inside to play piano with your toes. Steve Komito stresses, "A boot should feel good even when it's new. So often people are given the idea that they should wear a boot through suffering and pain and somewhere in the golden future it will feel good. I honestly don't feel that's necessary. The most critical point is adequate room in the toes, particularly for walking downhill. The boot has to keep your toes out of the toe box. The heel has to fit so there isn't excessive slipping. I suggest people lace up their boots snugly and walk. If there's heat after 15 or 20 minutes, there's friction, and if there's friction, there are going to be blisters."

SOCKS

Different sock combinations make a big difference in the way boots fit, so pick your socks before trying on boots. For cold weather, start with a thin synthetic sock. Then add a vapor-barrier sock, if you need one, then your insulating sock or socks. Here wool clings to its last foothold, so to speak, in my stock of outdoor gear, mostly because I haven't yet found a thick, compression-resistant, 100-percent-synthetic sock. The best wool and wool blend socks cost twice as much as the cheap ones and last three times as long. The cheap ones wear away to their nylon reinforcement faster than you can say fishnet underwear.

If you can find them, buy socks that come in sizes, without elastic, rather than stretch ones. Wearing too many stretch socks can constrict circulation and lead to cold feet. Stretch socks don't last as well, either, because they stretch most, and so are thinnest, at the greatest wear points: your heel and toes.

Some people prefer to wear two pairs of thinner socks outside the liner sock rather than one thick one. It's a little more difficult to get that many socks arranged on your feet without wrinkles, but the system does give you the option of removing a pair when your feet swell after half a day of hard walking.

Some climbers have experimented with using Neoprene wetsuit booties as socks. The closed-cell foam provides good insulation and also acts as a vapor barrier. That means that the same foot-care procedures I described earlier for cloth vapor-barrier socks are imperative with Neoprene socks. An additional danger may threaten high-altitude mountaineers using these socks. The gas inside the closed bubbles of the Neoprene can expand as the air pressure outside drops with increasing altitude. That could cause the socks themselves to expand and reduce circulation, increasing the risk of frostbite.

Putting on cold boots is often like putting your feet in a bucket of ice water: your feet get cold rather than your boots getting warm. Sleeping with your boots inside your sleeping bag beats the problem if you've got a big enough bag. Make sure you get all the snow off first so it won't melt and soak your bag, then put your boots in a stuff sack so they won't abrade your bag's delicate lining fabric. Wearing your boots day and night, without ever taking them off, is a good way to enrich the doctor who will soon be treating you for frostbite. Your boots, socks and feet will all need to dry out after a day of hard work. In addition, it's hard to get every last ball of snow off without taking your boots off. Any snow you miss melts and soaks your bag. My companions on Mt. Hunter's south face slept with their boots on for most of the climb. Toward the end, their bags had become nylon sheets enclosing clumps of frozen down— not a pleasant sight when you're facing a -20-degree night. Their damp socks provided little insulation. When we finally got off the climb after a grueling 13 days, both checked into Providence Hospital with frozen feet.

If you're going to be wading through deep snow, you need some way to keep snow out of your boot tops. The cheapest, lightest way is a pair of gaiters, simple fabric tubes that cover most of your boot and reach up to your knee. A strap under the arch of your foot holds them down; a zipper lets you put them on over your boots. That zipper takes a lot of abuse, so make sure it's substantial. Better gaiters have a storm flap that folds over the zipper and is held down with hook-and-loop. Snaps freeze shut fiercely in the damp environment where gaiters live. Avoid them.

Not even the best mountaineering and skiing double boots, as they come off the shelf, will keep your feet warm at high altitude when it's way below zero. In most boots you can put a closed-cell foam insole between the outer and inner boots. In some you can also replace the wimpy inner-boot insole with something thicker and warmer. Plastic double boots in particular have very thin soles, increasing heat loss through the bottom of your foot.

If you've still got cold feet after inserting insoles, you need to add insulation outside the boot.

Supergaiters cover the entire boot down to the welt, but leave the sole free. That's an advantage for mountaineers who want to climb difficult rock, and for skiers who need their boot soles free to get into their bindings. Most supergaiters come with foam or synthetic-fiber insulation. Many are designed so it's easy to add more insulation to them yourself. I improved my supergaiters by adding extra closed-cell foam around the foot and a layer of heavy pile around the calf. Adding insulation to your calf keeps the blood flowing toward your toes a little warmer. My feet got cold at 40 below at 17,000 feet, but didn't freeze.

The disadvantage of supergaiters is that they leave a large surface area, your boot sole, in contact with cold, conductive snow or even colder and more conductive crampons. Overboots enclose the entire boot, including the sole. That means you either climb rock with your crampons on, or take your overboots off. It also means you'll have trouble getting your boot attached to your crampons and skis. In some cases, you can modify the overboots to work. In other cases, particularly alpine ski bindings, you probably can't. But you do have the warmest boot setup you can get and still have a stiff boot somewhere inside. Few people really need overboots, so few companies make them. Try calling specialty mountaineering shops and asking them for sources.

If double boots with vapor-barrier socks, extra foam insoles and overboots don't keep your feet warm, you've either cold-damaged your feet somewhere along the line, or you aren't maintaining your body. Body maintenance, in fact, is just as important as gear in preventing frostbite. I'll talk more about that in an upcoming chapter.

· 6 ·
THE
OUTERMOST
NECESSITIES:
Gloves, Hats, Goggles
and Face Masks

GLOVES

My two friends and I knew we had to move fast to get up the south face of Alaska's Mt. Hunter with the limited food we could carry. We left behind everything we deemed non-essential. That included an extra pair of the nylon shell mittens called overmitts. After all, one brand-new pair should last for ten days, right? On the first day we began climbing snow-covered rock, which required tediously clearing each hold of snow. The constant abrasion began to wear on my overmitts. Swinging my ice axe for hours on end accelerated the deterioration. By the ninth day, gaping holes had appeared in the palms. Snow crept inside. Completely preoccupied by the struggle to ascend difficult ice in a 50-mph, subzero wind, I neglected my hands. By the day's end, all of my fingers and both thumbs had been severely frost-bitten. I ended up losing the tips of three fingers.

It doesn't take an Alaskan epic to destroy or lose a pair of overmitts. Lose control of one for just an instant in a high wind and it's gone. That's happened to me twice: once in a 100-mph gale in Rocky Mountain National Park, once in a much milder wind that blew the overmitt over a huge cornice and down a potential avalanche path.

The first principle of hand care, then, is to bring spare protection. That means not only spare overmitts, but spare gloves and mittens as well. I joke with my friends that the stuff sack containing my extra hand protection is as big as my sleeping bag, but I bring it nonetheless. Wintertime trekkers constantly handle snowy objects, which means gloves and mittens get wet quickly. Having spares makes all the difference in comfort. On long trips, I also carry yarn and a darning needle to patch thin spots in wool gloves and socks. I carry strong sewing thread and a thinner needle to stitch together small holes in lighter gloves.

To avoid losing overmitts, I often attach idiot strings. Although they make you feel like your mother just bundled you up to walk to grade school, they're very useful on steep climbs and in high winds. Some people run a string between the two overmitts, put on the overmitts, then put on their shell jacket so the string runs up one sleeve, across the back and down the other sleeve. That keeps the string out of your way, but holds the overmitts close to your wrists, which makes it cumbersome to reach into a pocket or your pack. Other people sew small elastic loops to their overmitts, which slip over their wrists. The same problem occurs. The best system I've seen is to tie a loop big enough for your head in the middle of a string connecting the overmitts. I put the loop over my head last, above all my garments including my hood. Then I put my pack on, followed by my overmitts. The crux is remembering to remove your overmitts and let them hang before taking your pack off. The system isn't perfect, because blowing snow tends to fill your overmitts when they're dangling by your knees. The only other system I've used is to keep some kind of spring-loaded clip attached to the D-rings commonly found on pack shoulder straps for just this kind of purpose. I snap a loop on my overmitts into the clip when the overmitts aren't in use. The clip I use is a carabiner, a three-inch long oval aluminum ring with a spring-loaded gate in one side. I use it because it's big enough to manipulate with gloves on. The clip-on system works well if you can take the time to clip

The best idiot-string arrangement has a long cord connecting the two overmitts with a loop tied in the middle to fit over the wearer's head.

and unclip your overmitts. With idiot strings you can jerk them off and let them fly like a kite in the wind if you're in a hurry.

It makes sense to layer your hands for the same reasons you layer your body: precise fine-tuning of insulation for different temperatures and extra warmth through air trapped between layers. With hand protection, there's an additional reason: layering means you can separate the layers for fast drying on long trips. Wet gloves and mittens won't dry if you stick them in your pack. You either have to put them deep inside your clothing and leave them there, preferably overnight, or conjure up a hot, sunny day.

In serious cold, I start my layering system with thin gloves that give me enough dexterity I almost never need to remove them. Polypropylene gloves work well, although they don't last too long. Polypro melts at a pretty low temperature, so if you handle hot pots while wearing them, you'll quickly get holes. Lightweight polyester gloves last longer. Wind blows right through both kinds, however, making them less than ideal for work outside the tent. Lately I've been using lightweight cross-country ski gloves, which are leather on the palm (including the palm side of my fingers), and knitted polypropylene on the back. The leather/polypro gloves are more resistant to getting wet than the all-polypro gloves, but take longer to dry. Leather resists heat and abrasion better than polypropylene. Before buying either kind (or any kind of sewn hand protection, for that matter), turn the gloves inside out and inspect the seams. Glove-makers use very narrow seam allowances, which means that a small sewing mistake creates a seam waiting to blow out. It's also a good idea to turn every sewn glove and overmitt inside out for an inspection before a long trip.

Although I'm a fan of vapor-barrier shirts and socks, I've only rarely used a vapor barrier on my hands. No one is now making a fabric VB glove. Lightweight polyethylene, PVC and natural-rubber gloves, available from surgical supply houses, last only a day or two. Kitchen dish-washing gloves always seem too tight unless I buy them so large my fingers look like E.T.'s Some people have used Neoprene skin-diver's gloves in wet, cold conditions on land and reported good success. I took loose-fitting rubber gloves when I went sea-kayaking along the coast of Alaska's Kenai Peninsula. The gloves kept the wind off my damp hands, so my hands felt fairly warm, but they also gave my skin the look of a baked apple.

Pogies are another solution for cold-water paddling. These pile-lined paddling mitts wrap around the paddle shaft and attach with hook-and-loop. Your hand grasps the shaft directly, without layers of pile and nylon in between. Water sneaks in eventually, but they're considerably better than nothing.

In moderate cold, down to 10 or 15 above as my hands measure temperature, all I need above my liner gloves is a pair of insulated overmitts. If it's really cold, however, I substitute a pair of very heavy wool gloves for the liner gloves. I'd choose an all-synthetic glove or a wool/synthetic blend if I could find gloves that were thick enough. The synthetic would probably last longer and dry faster. As a backup for the wool gloves on extended trips, I carry a pair of thick wool mittens. Mittens are inherently warmer than gloves because they have less surface area in proportion to their volume. Every square inch of surface provides an opportunity for heat loss by radiation, convection and conduction. Minimizing surface area also minimizes heat loss.

All kinds of knitted gloves and mittens should be

protected from snow. I've heard people argue that the
snow simply forms a crusty layer on the surface without
affecting the insulation. Don't be deceived. Enough
snow melts to saturate the wool, which then loses its
insulating capacity and takes forever to dry.

The insulated leather gloves used by downhill skiers
seem at first glance like an ideal combination of insulation and a water-resistant shell. They do work well for a
day on the slopes—if you can dry them at night in the
lodge. In the field they dry very slowly because on most
of them you can't remove the insulating liner. They also
lack a gauntlet, a long cuff that reaches up your fore-
arm and closes the gap between your shell jacket's cuff
and your hand protection. A friend once had his wrist
severely frostbitten while ice-climbing on Huntington's
Ravine on Mt. Washington simply because his hand
protection didn't include an adequate gauntlet.

Overmitts, by definition, include a gauntlet. Look for
tough fabrics, particularly on the palm and in the notch
between your thumb and first finger. Look, too, for clo-
sures at the wrist and gauntlet cuff that are easy to adjust, even with an overmitt on the other hand. If you
leave the gauntlet cuff open, heat slips out and snow
sneaks in. Some overmitts come with sewn-in, insulating liners. I prefer the kind with removable pile liners,
usually held in with hook-and-loop, because they can
be taken out for rapid drying.

Before you actually buy all the separate pieces of
your hand layering system, make sure they fit together.
Some insulated overmitts will only accommodate a thin
liner glove inside. Trying to fit in a heavy wool glove or
mitten can force you to buy a larger size—which sometimes means the overmitt's finger section is way too
long. Often that problem can be corrected with a sewing machine and a good pair of scissors.

HATS

Layering applies as much to your head as to your hands. For starters, it's a good idea to have some way to cover your neck, which is just as well supplied with blood as your head and loses heat just as fast. A scarf works fine for sleigh rides on snowy evenings, but it's more cumbersome to use in the woods than a balaclava.

Balaclavas are hats that cover everything from your shoulders up except for a circle surrounding your eyes, nose and mouth. They were named after Balaklava, the Russian seaport on the Black Sea where the English cavalry made the ill-fated attack celebrated in "The Charge of the Light Brigade." Some balaclavas are made of very tight, stretchy silk or polypropylene. With these models, however, I usually find myself chewing on the balaclava at mealtimes instead of my food, so I prefer looser, thicker models. My favorite long-under-wear top has a polypropylene balaclava sewn to it. I sewed the balaclava on to keep it handy and prevent it from rotating around my head and blinding me. I added an 18-inch zipper from chin to sternum for ventilation. In moderate cold I wear a knitted ski hat over the balaclava. When it's wicked cold I wear a thick pile balaclava over the polypropylene one. Wool offers no advantages in headwear. It just dries slowly and itches.

GOGGLES AND FACE MASKS

A balaclava alone will not keep your face from freezing when the wind is sandpapering your skin with blowing snow. A good pair of ski goggles helps a great deal. Buy the double-pane variety. Although they are usually more fragile, they ice up less from condensation inside than the single-pane ones.

Goggles still leave the tip of your nose, chin and lower cheeks exposed. For those really ugly days when you can't even look into the wind and secretly wish you were windsurfing in Tahiti, you need a face mask. I like the Neoprene variety, which fits neatly under the lower edge of my goggles and covers every last scrap of flesh left exposed by my balaclava. Neoprene face masks stop the wind completely, unlike the thin knitted "face masks" which are actually shaped like balaclavas. Breathing through a fabric face mask quickly causes it to ice up. Face masks made of eighth-inch thick Neoprene offer enough insulation to ice up less.

With a balaclava, hood, goggles and face mask, you can stare the worst snow-laden gale in the face and laugh—so long as it doesn't simply knock you flat, that is. I had always regarded face masks as too claustrophobic for extended use until three friends and I embarked on a 100-mile ski tour from the old ski resort of Eldora, in Colorado's Front Range, to Vail. The first day we spent five hours above timberline on the Continental Divide, struggling through a horrendous gale. Without full face armor, we would either have turned back or risked freezing our cheeks. With it, we were able to find our way by map, compass and altimeter through the blinding storm. Six days later, we arrived in Vail.

· 7 ·
EIGHT INCHES OF HEAVEN:

How to Sleep Comfortably in the Cold

DOWN VS. SYNTHETIC

For three days, George Lowe, Lito Tejada-Flores and Chris Jones had struggled with the steep rock on the south face of Devils Thumb, in the Coast Range near Petersburg, Alaska. Then the weather went sour. The storm began as cold rain. As the temperature dropped, the rain turned to sleet, then snow. Ice coated everything, including their garments. The loft of Tejada-Flores' down jacket shriveled to nothing as water saturated the down. Lowe remembers stuffing his fiberfill sleeping bag in the morning and seeing water ooze out. When he pulled it out in the evening, he could feel clumps of ice. Nonetheless, he stayed warm—at least, warm enough to survive.

"Without fiberfill bags, we simply couldn't have done it, or we would have died," Lowe says. "It's important for people to realize that in that climate, down is useless."

In any climate, down sleeping bags require more care than synthetic ones. In some situations, down bags can be positively dangerous. Lowe and his friends had tackled a 2,000-foot rock wall in storm-wracked coastal Alaska. Their bivouac ledges, at times, could barely accommodate both buttocks. Pitching a tent was impossible. Inevitably, their sleeping bags got wet.

I count myself lucky that I've never been caught on a big Alaskan rock face in such utterly miserable conditions. My mountaineering epics have been in colder, drier conditions on higher peaks. Naturally, I needed a much warmer bag. I chose a down bag because it weighed at least a pound less than a synthetic one of similar warmth. When starting on a ten-day alpine-style ascent of McKinley, weight is critical. Had conditions been dry but milder, the difference in weight between suitable down and synthetic bags would have been somewhat less. A suitable down bag for McKinley might weigh four and a half pounds, a suitable synthetic, five and a half—a pound difference. A suitable down bag for the Sierras might weigh three and a half pounds, a suitable synthetic, four and a quarter—a three-quarter-pound difference.

To keep our down bags dry on long climbs in Alaska, we would get up at 3 or 4 a.m., climb until 3 p.m., then camp. Our yellow tent would become a greenhouse, drying our bags completely so we could sleep warm during the −20 to −40 degree night that would follow. The habit of drying his bag whenever possible became so ingrained in Peter Metcalf that the first thing he did after we checked into a motel following our Mt. Hunter climb was yank out his damp bag.

Synthetic bags should be dried whenever possible, too. Du Pont's data shows that the insulation of a Quallofil bag drops 15 percent when the bag has absorbed 25 percent of its weight in water. A down bag's insulation drops 50 percent.

Backpackers in less sodden climates than coastal Alaska can certainly remain comfortable in a down bag if they keep their down dry. As I detailed earlier, good down still offers more loft for its weight than any of the high-loft synthetic fills. Down also provides roughly 20 or 30 percent more insulation for each unit of *thickness,* probably because it slows the escape of infrared radiation a little better. Kansas State University's Environmental Research Laboratory tested five down and synthetic bags of different weights. The 5.5-pound down bag gave 7.65 clo, one clo more than the 5.5-pound synthetic bag. Looked at another way, the 4.5-pound down bag gave as much insulation as the 5.5-pound synthetic one.

If you examine two bags from the same manufacturer with the same comfort rating and similar lofts, one filled with down, one filled with a synthetic, you'll find a 20 to 30 percent difference in weight. Du Pont, measuring sections of sleeping bags, puts the difference between finished Quallofil and down sleeping bags at about 15 percent.

Fiber manufacturers constantly improve their products. The time may come when a synthetic can match down's warmth-for-weight, compressibility and longevity. When it does, the synthetic will probably become the clear choice. Until that time, the choice between synthetics and down is a trade-off that depends on the conditions you'll encounter, the importance to you of weight and stuffability, how much energy you want to put into keeping your gear dry and your pocketbook.

Down bags, like down parkas, cost much more than synthetic ones.

If you choose to go the synthetic route, your choices now in quality bags are Quallofil, Polarguard and Hollofil. Quallofil gives a bit more warmth for its weight and better compressibility. PolarGuard and Hollofil are in a two-way tie for second. Bag makers vary in how they use the products, however, and the products keep evolving, so you'll have to do some research to decide which synthetic is actually best when you're ready to buy. Look at the manufacturers' claims, then carefully examine the finished bags in the shop. Lay them out on the floor and measure the loft. Compare that to their weight. Roomier bags, of course, will weigh more. Stuff the bags in their stuff sacks. Bags with the same filling can differ a good deal in loft for weight and stuffability simply because of different construction techniques.

If you choose down, you need to consider down quality. Down fill power is measured by putting one ounce of down into a graduated cylinder and dropping in a light weight called a platen. The down's fill power is the volume of the cylinder below the platen, measured in cubic inches. Price, compressibility and loft for weight are all proportional to fill power. Pretty good down lofts 550 cubic inches per ounce. Excellent down averages 625. According to some reports, down from the eider duck can loft 1,200 cubic inches. That gossamer stuff comes only from far northern Europe and Iceland, where only a few thousand pounds are collected each year from the nests of wild eider ducks. Eider down costs something like $600 a pound at wholesale. A simple eider down comforter will set you back a couple of grand.

Returning now from the stratosphere, how much difference does down quality make? If you made a bag with 40 ounces of 625-fill down, you'd get a really warm winter bag with about nine inches or so of loft. An equally lofty bag made with 550-fill would weigh about five and a half ounces more and take up five or ten percent more space when stuffed. If you're intent on cutting ounces and cubic inches before embarking on some desperate climb, that might be significant to you. A better reason to seek out high-quality down is that any company using such a premium product is probably building excellent bags. If the lofts are equal, the difference in warmth between a 625-fill bag and a 550-fill one is probably negligible. Durability should be comparable, too.

By FTC regulation, a sleeping bag advertised as containing "100 percent down" must indeed contain nothing but down. However, a bag labeled simply as containing "down" can have up to 20 percent down fiber (barbs that have become detached from down), waterfowl feather fiber, and waterfowl feathers. Feathers, in particular, lack the resiliency, loftiness and durability of down. To get some sense of the feather content, feel the bag. A high feather content makes a bag feel stiff, as if it contained straw, when compared with a bag containing better down.

Labels like "northern," "prime," "white" and "gray" lack any legal definition. They mean about as much as the label "new and improved" on a laundry detergent. By law, the label "goose down" must mean that 90 percent of the down is from that species. Goose down tends to loft better than duck down because the larger bird has larger and more lofty down plumules. However, poor goose down lofts less than good duck down. Most down sold today is a mixture. Fill power per ounce indicates quality better than species. Many of the better manufacturers test their own down regularly— another good assurance of quality.

BAG SHAPE AND FIT

After picking the fill, think about overall shape. Rectangular bags are roomier than tapered, form-fitting, mummy bags, so they feel less confining, but they offer much less warmth for their weight. First, you've got the extra shell and filling to carry around. Second, you've got larger pockets of air around you. When you roll over, you tend to expel that warm air through the bag's entrance, then suck cold air back in. Third, the heat loss from a bag is related to its surface area. Rectangular bags have more surface area than mummy bags, which means a higher rate of heat loss. Since your heat output is nearly constant when you're asleep, that means you need a thicker bag if it's rectangular than if it's a mummy. Nearly all winter bags intended for *your* back, not a horse's, are mummies.

Ultimately lightweight, efficient bags have room for you and nothing else. In severe cold, however, you'll want to sleep with your water bottle (if you fill it in the evening to save time in the morning), and possibly your boots and camera. In addition, if the bag's too snug, you won't be able to sleep inside with enough clothing on. You don't need to leave room for your parka, however. Since the portion of the parka that's under your back gets squashed flat, you get more benefit from your parka by laying it unzipped on top of your bag and holding it in place with tabs of glue-on hook-and-loop. That way none of the parka is compressed.

The only way to find out if a bag fits is to shuck your shoes and crawl inside with all the gear you intend to sleep with. Remember that you not only have to sleep in the bag, you may have to live in it for a few days if you get storm-bound on your next traverse of the Patagonian icecap. Decide if it fits with that in mind.

How much loft you need depends on your metabolism and how willing you are to sleep in lots of clothes. Manufacturer's temperature ratings are useful mostly for comparing bags within one bag maker's line, not for comparisons between manufacturers. I've taken a bag rated to −5 degrees F and slept in it, passably comfortable, when it was −33 degrees. However, I was using a vapor-barrier sleeping bag liner and wearing every scrap of clothing I owned, including four hoods and hats. My friend, sleeping beside me in an identical bag with similar amounts of clothing, shivered half the night even when the temperature was 25 degrees warmer. The best you can do for your first bag is to take the manufacturer's comfort rating as a basic guideline. If in doubt, go a little on the thick side. When you're tired, damp and underfed and your bag hasn't seen the sun in a week, you'll be glad you did.

BAG CONSTRUCTION

The method of putting synthetic fibers into a sleeping bag has little effect on warmth per unit of loft, but it does affect loft for weight. Continuous-filament battings like PolarGuard are usually sprayed with a light resin which, along with the long, intertwined filaments themselves, holds the batting together. Stitching along the batting's edges holds it in place within the bag. Staple-fiber battings can also be resinated and stitched in. In most cases, resinating a batting reduces its loft and compressibility but can increase its strength and durability.

Staple-fiber battings can also be made as a sandwich with a nonwoven fabric called a scrim as the bread. Battings made with a scrim must have smaller compartments between stitch lines to prevent shifting during use or washing. Battings that shift have uninsulated cold spots.

Synthetic battings can be laid flat, one on top of the other, with the quilt lines offset, or they can be stacked like shingles on a roof, with each layer overlapping its neighbor. Both methods, if done well, can allow the fiber to loft to its full potential.

Down, like staple fibers, must be confined in compartments to prevent shifting. The fabric walls that stretch between the inner and outer shells of the bag to form those compartments are called baffles. The stitch lines you see running at right angles to the bag's long axis are where the cross baffles are attached. Side baffles or channel-block baffles run the length of the bag along the side opposite the zipper.

Far and away the most common down construction method uses slant-wall baffles, in which the cross baffles lean toward one end of the bag like dominoes in the process of falling. Theoretically, that prevents the stitch lines attaching the baffle to the inner and outer shell from being directly above each other, so the bag is supposed to be warmer. In truth, a well-filled bag won't have cold spots regardless of the baffle angle. Down in an underfilled bag, regardless of design, will tend to fall off to the sides as you sleep, leaving only a thin top above you. To check, grab one side of the bag and give it a gentle shake. If the down stacks up on the far side, causing a difference in loft across the width of the bag, the bag may be underfilled. Comparing good bags to poorer bags is the easiest way to see the difference.

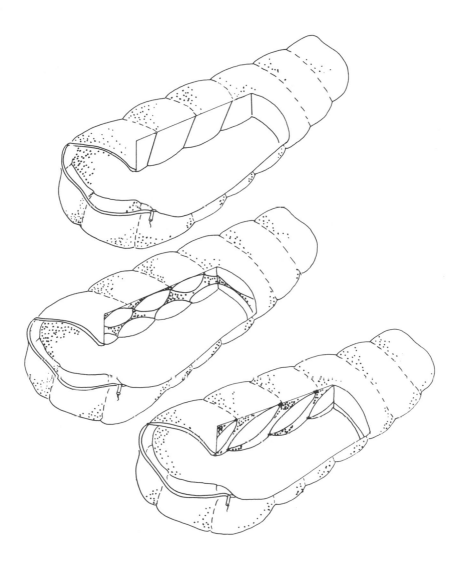

Sleeping bag construction. Top: *Slant-wall baffle construction, used in down sleeping bags.* Middle: *Overlapping quilt construction, used in synthetic sleeping bags.* Bottom: *Shingle construction, used in synthetic sleeping bags.*

One fine point to look for in a down bag is channels that are fully closed at the side of the bag away from the zipper. That prevents down from shifting over time from channel to channel, eventually accumulating in one place, leaving the rest of you to shiver. To check, press a thumb into the junction between a cross baffle and the side baffle. If your thumb goes through, no seam exists.

Another feature to look for is a generously stuffed draft tube lying against the zipper to prevent cold air from entering there. The best bags have two draft tubes. Zippers should be sturdy so you don't blow one out on a −20-degree night. The fabric right next to the zipper should be backed with a stiffener so the zipper slider doesn't snag the fabric and hang up.

Quality bags have foot sections cut so the inner shell is smaller than the outer one. That helps prevent your feet from pushing the inner shell into contact with the outer shell, shoving aside the insulation in the process.

Most critical of all, well-made bags have hoods that seal snugly around your face. The best bags have both an insulated collar that closes around your neck with its own drawstring, and a second, conventional hood that closes around your face. That system blocks the escape of warm air better than a conventional hood alone. It also lets you cinch down the collar while leaving the main portion of the hood open. That helps prevent the moisture in your breath from penetrating the fabric and condensing in the insulation. In many good down bags, the inside portion of the hood is made of a waterproof-breathable fabric, which also helps keep the hood's down dry.

Besides drawstrings, hoods usually have hook-and-loop tabs to fasten them securely after the zipper is closed. Those tabs should be easy to find in the dark and at least an inch wide for good holding power. I've seen systems so complicated that you need your head-lamp on and both arms out of your bag to fasten the hood properly. Once fastened, of course, you can't get your arms inside the bag without unfastening it again. The half-inch wide tabs found on some bags pop open at a sneeze.

Regardless of hood style, don't cinch down the outer hood with your head entirely inside your bag. Your breath will condense and saturate your insulation. With a down bag, you're likely to start losing loft after a few nights unless you can dry it. With either synthetics or down, you're likely to wake up in the middle of the night with some pretty cold, moist fabric caressing your cheeks. Even with the outer hood cinched down loosely you're likely to get some frost forming on the hood if it's cold enough. To shield my skin from direct contact with that frost, I often wear a lightweight poly-propylene balaclava to bed. If you find yourself con-stantly tempted to put your head entirely inside the bag, consider getting something warmer.

BAG CARE

Clean bags insulate much better than dirty ones. Dirt can cause down and polyester fibers to clump together, which reduces loft. To clean a bag, hand wash it or use a large-capacity, front-loading washer. Top-loading, ag-itator-type machines are far too rough on bags. Down is heavy when wet and agitation tends to rip fragile baffles loose. Fiberfill battings can develop thin spots. Use a mild soap that dissolves easily in the local water, then rinse thoroughly. Soap residues kill loft worse than dirt. Take care that the wash-water temperature doesn't exceed 140 degrees. At least some polyester fibers begin to lose their crimp at that temperature, which leads to loss of loft. Never stuff either kind of bag, then leave it some place hot, like the trunk of your car in the summer. Don't wring out a wet bag. Instead, roll the bag in several towels and gently press out the remaining water. Air dry or tumble dry on air-only setting. You can start the drying process on low heat, but you should finish drying on air-only or outside the dryer. Many home dryers can get the outside of the bag to 165 degrees F, even set on low. No matter what drying method you use, you'll probably have to do a lot of fluffing to restore a down bag's full loft.

Don't dry-clean any kind of bag. Some manufacturers assert that dry cleaning strips down of its natural oils, rendering it brittle. Whether that's true or not, dry cleaning can leave a residue in the bag if the rinse solvent isn't clean enough. Like a soap residue, that can hurt loft. And fumes from chemical residues can make you think you're sleeping alongside the New Jersey Turnpike instead of deep in the piney woods.

SHELLS, LINERS, AND AIR MATTRESSES

Bags with waterproof-breathable outer shells are a blessing on extended trips. Snow caves and igloos can be drippy. Frost forms on tent liners, then sifts down gently when the wind rattles the tent. Spindrift, fine snow carried by the wind, can eddy through snow-cave entrances and tent doors and vents and fall onto your bag. Waterproof-breathable fabrics keep that moisture out of your bag. That's particularly important with down bags, but not trivial even with synthetics. An alternative is a waterproof-breathable sleeping bag cover.

It's just as important to keep moisture from entering

your bag from the inside. That means using a vapor-barrier liner. On long trips, I start using the vapor barrier as soon as I land on the glacier just to keep my bag dry, even if I don't need the warmth.

Radiant-heat-barrier (RHB) liners can also add significant warmth. Some double as vapor barriers; others have tiny perforations intended to permit the escape of some sweat vapor. Some RHB liners are made of an aluminized film that looks like a lamination of aluminum foil and plastic sandwich wrap. Others are made of aluminized fabric. Tests at Kansas State University showed that adding a reflective layer to a synthetic bag weighing three pounds ten ounces increased the clo value by 14 percent while only adding six percent to the weight. Radiant heat barriers offer less benefit in down bags because down by itself blocks radiant heat better than the synthetics.

Your body weight will completely flatten any sleeping bag insulator. To avoid freezing to death through conduction into the cold ground, you need compression-resistant insulation beneath you.

Open-cell foams compress readily enough that they need to be fairly thick, at least one and a half inches, to keep you off the ground. That means they're heavy and bulky compared to closed-cell foam. Like a kitchen sponge, they also absorb water, so a waterproof cover is essential. Their insulation value when compressed is good if you've still got enough thickness left.

Closed-cell foams give very little under body weight, so they can be used much thinner. They're also less comfortable. Three-eighth and half-inch thicknesses are standard. A three-eighths-inch closed-cell foam pad gives roughly 1.4 clo. Your winter-weight sleeping bag with six to eight inches of loft still provides three or four inches of insulation above you even when the bottom half is squashed flat. That means you've got eight to ten clo of insulation above you and one-sixth of that or less below if you're using one three-eighths-inch pad. Furthermore, your body weight is pressing the pad into contact with cold, highly conductive snow, ice and frozen ground, while the top surface of your bag is only in contact with air. Obviously, one pad alone doesn't do the job in severe cold. You'll find yourself rolling over constantly, trying to rewarm your cold underside. Solution: carry two pads. Glue them together at one end if you don't want them to shift around. Just don't glue them together everywhere or you won't be able to roll them neatly. Beware of pads that stiffen in the cold. A friend once tried to roll up a summer pad on a cold winter's morning and ended up with something resembling a pile of fence slats.

Air mattresses keep you off the cold ground, but the empty space between the top and bottom of the air mattress is large enough that spontaneous convection currents form, draining away heat. If you like an air mattress' comfort and don't mind the weight, add a closed-cell foam pad on top.

Some air mattresses have open-cell foam inside, which stops the convective heat-loss problem. They're luxurious to sleep on, but, like all air mattresses, have an Achilles heel, as a friend of mine demonstrated by playing mumblety-peg near his mattress at the start of a three-week McKinley expedition. A mistimed throw of his knife neatly skewered his mattress. Lacking a patch kit, he was literally flat on a very cold back for the rest of the expedition. The moral, of course, is to bring a patch kit. Then hope the valve doesn't fail, which would cause an equally catastrophic, but probably unrepairable, breakdown.

·8·
HOME IN A STUFF SACK:
Tents for Winter Camping

At dawn the clouds lifted, but the inhabitants of the high camp at 17,200 feet on Mt. McKinley weren't about to celebrate. With clearing came wind: 60, 70, 80 miles per hour and beyond. At 6 a.m., National Park Service Ranger Roger Robinson crawled out of his beleaguered tent and dashed for his neighbor's. Inside, guide Nick Parker and an assistant huddled against the tent wall on the windward side, trying to keep the poles from snapping like dry spaghetti noodles.

"We've got to dig a snowcave!"Robinson screamed. Parker nodded. Robinson slipped out of the tent and scuttled crab-like across the snow to check on other people.

Climbers everywhere were struggling with their tents. Now and then a foam pad, a sleeping bag—once even a wallet with $500 cash—escaped a numb, mittened hand and vanished into the abyss to the north. Then Robinson saw two fully erect domes dragging Parker on his face across the snow. Someone had inadvertently removed the last ice axe holding them down. Parker had lunged for them, caught hold of one with each hand, then discovered too late that not even his full body weight could restrain them. The runaway tents dragged him closer and closer to the drop-off at the edge of camp. At the last minute he simply let go and watched both tents sail into the blue.

When the wind finally died, only one tent out of ten remained erect. Credit for its survival belonged more to the four-foot-high snow walls surrounding it than to the tent itself.

No tent light enough to carry comfortably on your back will stand up forever to drifting snow and high winds. Himalayan climbers sometimes use a shelter called a Whillans Box, but as Everest veteran Todd Bibler said, "That's not a tent—that's a building." A Whillans Box weighs 30 pounds. If the weather really gets extreme, it's time to start thinking about a snow cave. Even if your tent can withstand the wind, your nerves probably won't. Jack Stephenson tells the story of a customer named Larry who took one of Stephenson's extra-rugged, three-person tents to the summit of Mt. Washington in search of a 150-mph gale. Larry had an anemometer that read to 90 mph. The winds blasted the needle off the scale. The summit observatory recorded winds between 120 and 132 mph. Larry couldn't stand up in the wind, much less sleep in the thrashing tent. Although the tent survived, Larry decided afterwards that he really wasn't interested in sitting through a 150-mph wind.

In more sane conditions, tents have many advantages over snow caves. Digging a cave sucks up a lot more energy and time than pitching a tent. Caves can't be dug everywhere, either. In good weather, living in a snow cave is like living in a freezer, while tents warm up beautifully.

The simplest "tent" is nothing more than a waterproof fabric sack, called a bivy sack. Some come with flexible wands that support the fabric around your face. They're big enough to sleep in, but not big enough to cook, dress, or even play solitaire inside. They're better than nothing, however, as I learned the hard way on my first major winter climb, in Rocky Mountain National Park. My partner and I took only sleeping bags, thinking that the uncoated nylon shells would be enough to keep the snow off. By 2 a.m., spindrift sifting down the gully where we'd bivouacked had buried us completely. Melting snow had soaked our bags and begun chilling us rapidly. Unable to sleep, we dug ourselves out and started climbing again by headlamp. A bivy sack would at least have kept us dry and warm. If we had brought a larger shelter, half of it would have overhung the edge of the tiny platform we'd been able to chop.

Let's say your thirst for winter adventure is satisfied by more normal pursuits, such as overnight ski tours and the like. For that, you not only need a tent, you can use one.

Most people think a tent should keep them dry, period. In truth, that task is partly the tent's job, and partly yours. Let's talk about the tent's job first.

TENT DESIGNS

The most common backpacking tent design keeps rain and snow out with a waterproof, coated fly that is stretched over an uncoated canopy. If the fly extends low enough, it can easily prevent wind-driven horizontal rain and rain splashing up from the ground from entering the tent. A more difficult problem, under some conditions, is getting wet from condensation inside the tent walls.

The water vapor in any sample of air, if cooled enough, will condense to water droplets. The temperature at which condensation begins is called the dew point. If the tent fabric's temperature is below the dew point, water vapor will condense on it. A single-wall, unventilated (note that word) waterproof tent would generate horrendous condensation. Standard design "solves" that problem by introducing a second layer of uncoated fabric inside the waterproof layer. In theory, water vapor will pass through the uncoated canopy fabric and condense on the fly. Water droplets will run down the fly and drip onto the ground; frost on the fly will be shaken off by tent movement, then fall onto the tent canopy and slide down to the snow.

This design has two problems. First, if time or weather prevent drying that wet fly, it must be packed wet. Second, water vapor doesn't always pass through the uncoated canopy. In some temperature/humidity combinations, both above and below freezing, the temperature of the canopy can drop below the dew point, which means water vapor condenses right there. If it's calm, everything touching the tent wall gets wet. If it's breezy, inhabitants must endure a rainstorm or blizzard inside their tent. Severe condensation can occur even if it's clear outside. Standard tent design by itself doesn't always solve the condensation problem. For the most part, it simply tries to hide it.

For a while, standard design had two competitors. Some companies were making single-wall tents out of Gore-tex. The single wall made them very light and compact. However, Gore-tex breathes less than most of the porous fabrics used in tent canopies. Predictably, if the Gore-tex temperature dropped below the dew point, water vapor condensed immediately on the inside of the fabric. In some cases, condensation was worse in a Gore-tex tent than in a standard tent because the standard double-wall design gave a little insulation to the canopy. That kept the canopy slightly warmer, sometimes raising its temperature above the dew point. W.L. Gore added a non-woven polyester fabric called Nexus to the inside of the Gore-tex laminate to prevent frost or moisture from falling onto the occupants. The Nexus acted like an old-fashioned cotton frost liner to absorb and hold the condensation until the occupants got a chance to dry the tent. When the sun came out the Nexus was supposed to prevent dripping until the moisture could sublime through the Gore-tex film. The theory worked if the storm was not prolonged. During lengthy storms, Gore-tex tents tended to get wet and heavy and stay that way. Gore-tex alone didn't solve the condensation problem either.

In January, 1986, W.L. Gore took its tent laminate off the market because it couldn't meet fire-resistance standards. The market exists, however, so Gore or someone else is likely to put a fire-resistant, waterproof-breathable tent fabric on the market one of these days. You might also run into someone selling one of these tents used.

At first glance, the third basic tent design looks like it would generate the worst condensation. In reality, it may well cause the least. This design calls for a double-walled tent made exclusively from coated fabrics. The coating on the canopy prevents warm, moist air from reaching the cold fly, so the fly stays drier. The canopy coating also helps keep warm air in the tent, which keeps the canopy warmer and thus more often above the dew point. To keep the canopy warmer still, one company uses an aluminized ripstop to reduce radiant heat loss. I've used tents with this design for a total of 85 days in the Alaska Range. Compared to standard design, condensation was always minimal.

VENTILATION AND MOISTURE

Not even double-wall, fully waterproof tents can solve the condensation problem entirely on their own. The ventilation system is just as important. The key to successful ventilation is the chimney effect. Warm air rises, whether it's heated by the fire in your fireplace, your body or your stove. To let it escape, there must be a vent near the tent peak. Air can't flow both in and out of one vent at the same time, so there must be another vent where cold air enters to replace the warm. Logically, that vent should be near the floor. Tents with vents at only one level, be it high or low, do not ventilate nearly as well as tents with both high and low vents. With proper venting, even waterproof-breatheable and traditional tents can stay reasonably dry. Vents should be covered with very fine no-see-um netting, which effectively stops most wind-driven snow. Zippered covers over the vents prevent snow from entering in really nasty conditions.

Leaky tent floors are another source of soggy gear, clothing and morale. The tighter the floor fabric weave, as expressed by the number of threads per inch in each direction, and the heavier the coating, the more waterproof—and heavier—the floor.

Urethane-coated fabrics rarely block water as effectively as a sheet of plastic. Water vapor can penetrate some lightly coated tent floors and condense under your sleeping pad. Liquid water can penetrate seams that pierce the floor. A "bathtub floor," in which the floor/sidewall seams are raised above the ground, is desirable but not a complete solution. The best procedure when camping on soggy ground or wet snow is to use a three-mil plastic ground sheet. Cut the ground sheet to the tent's exact shape so rain won't pool on it, flow under the tent and penetrate the floor. Ground sheets reduce the need for a heavy tent floor. If it's really cold, you can get away without a ground sheet even with a very lightweight tent floor.

Doors can be another leak point. Tunnel entrances prevent snow and rain from entering when you're exiting, but they also turn your exit into an awkward re-enactment of the birth experience. Zippered entrances set into a sloping surface can be a problem if snow builds up on the tent wall. Opening the door tends to allow snow to pour in. Doors that flip out sideways toss any snow buildup away from the interior. Doors that fold down when unzipped drop snow buildup right into the tent unless you knock the snow off carefully beforehand.

Using a ground sheet and taking care opening the tent door are just two of the ways you can do your part to prevent your dry, snug tent from becoming a swimming pool. In addition, you should seal the tent's seams with a commercial seam sealant, then renew the coating regularly. You should also minimize the amount of moisture you bring inside. Bringing clothing soaked with rain or sweat inside to dry invites condensation. Your body and stove will warm the air to a temperature higher than outside, causing moisture to evaporate from your wet clothing. The tent wall, however, is cooled by contact with the outside air. When the humid air inside the tent hits the cold tent wall, condensation is likely. Shed your rain gear before getting into the tent, then turn the garments inside out so the moisture won't evaporate off them. If the walls do get wet, wipe them down with a sponge. If they get frosty, pile your gear in a corner and scrape them off. Then use the small whisk broom you remembered to bring to sweep up the frost as well as any snow that's blown in and toss it out the door.

Cooking in a tent adds a lot of moisture. I've seen sheets of ice one-sixteenth of an inch thick covering tent walls after several days of cooking inside during constantly stormy weather. Getting the frozen zippers open after a meal took a cigarette lighter and extreme care. And this happened inside one of the best-ventilated tents around.

The only real solution is to cook outside. In stormy weather, however, that's miserable at best, impossible at worst. If you must cook inside, cook directly under a vent so steam escapes immediately. Keep the pot covered as much as possible. If you have the fuel, run the stove for several minutes after you're through cooking to help dry the tent.

Never prime a white gas stove with white gas inside a tent. The danger of fire is just too great. Light the stove outside, and bring it in only when the flame has stabilized. Better yet, use a compressed-butane stove that doesn't require priming. Tents are fire-resistant, meaning they won't continue to burn after the ignition source is put out. But they will certainly burn if you've got a flaming pool of gasoline on your tent floor!

Ventilation when cooking inside a tent is more than a convenience. It could save your life. Two very experienced mountaineers died on McKinley in 1986 from carbon monoxide poisoning caused by cooking inside a sealed tent. Tired from a hard day, they apparently fell asleep while the stove was running and never woke up.

TENT SHAPE AND CONSTRUCTION

The driest tent isn't worth its weight in shredded nylon if it won't stand up in a storm. Tents suitable for winter come in two basic shapes. Dome tents look just like the name implies. Their interlocking pole structure gives them great strength and a minimum amount of unsupported fabric. All those poles do increase the weight, however. Two-person domes typically range from seven to nine pounds. They have about 35 percent more volume than an A-frame tent of equivalent floor area and height.

A second popular design for winter tents is the hoop or "Conestoga wagon." This design resembles the traditional A-frame except that the poles are inverted U's instead of inverted V's. The design's advantage is reduced weight compared to a dome tent (because of fewer poles) and about 58 percent more volume than an A-frame of equal height and floor area. Two-person hoop tents run from three to six pounds. Their disadvantage is that less of the tent fabric is directly supported by the poles. However, well-made hoop tents pitch skin-taut. Tents that don't flap rarely rip from wind load alone.

Traditional A-frame and single-pole pyramid tents have about vanished from the packs of serious winter campers. Though sturdy, the amount of usable volume per square foot of floor area is small because of the uniformly sloping tent walls. Hoop and dome tents have nearly vertical walls near the floor. Another drawback is that A-frames and pyramids usually require large numbers of stakes. Domes are free-standing, but must be staked to keep them from blowing away. They're also stronger if the corners are nailed down. Hoop tents typically only need one or two stakes at each end.

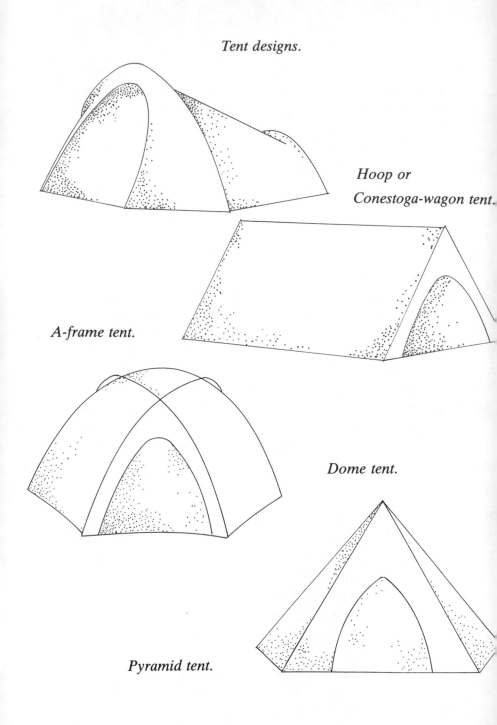

Tent designs.

Hoop or
Conestoga-wagon tent.

A-frame tent.

Dome tent.

Pyramid tent.

In the tent failures I've seen, the poles snapped first, then the jagged ends tore the fabric. The tensile strength of the pole material, the diameter of the tubing and its wall thickness determine pole strength. A further consideration is the amount a normally straight pole is curved when you pitch the tent. Simply erecting the tent often places the outside surface of the pole under a strain of approximately 45,000 pounds per square inch—about half the total tensile strength available. Not only does this reduce the amount of strength available to resist wind and snow loading, it introduces another problem: stress-corrosion cracking. Poles stressed over time, especially if the protective anodizing has worn off, can fail suddenly even without additional loading. This is why tent makers recommend you keep your poles clean. Dirty poles corrode faster. A short length of copper tubing with an inside diameter slightly larger than the outside diameter of the pole can be used to splice a break.

Manufacturers use flexible poles to keep the canopy taut. The remaining pole strength of 45,000 psi for a typical high-quality tent pole is sufficient for reasonably heavy loads. A few companies use permanently curved poles so that the full strength of the material is available to resist stress when the tent is pitched. Expensive tents usually have higher-quality poles.

Some tents come with a choice of fiberglass or aluminum poles. While recent improvements in fiberglass have increased its strength, it's still inherently weaker and heavier than the best aluminum. It's also cheaper.

Both kinds of poles are usually held together with elastic shock cord. In severe cold, the shock cord can lose its elasticity. That's not a problem taking the tent down. However, reassembling the poles can be difficult because the stretched-out shock cord won't contract and fit back inside the poles. The solution is to rapidly and repeatedly stretch and release the cord. That warms it enough to regain its elasticity and slip back inside the pole.

Even the best poles eventually fatigue after enough flexing in punishing winds. Eventually, they will break, sometimes under a lower load than they withstood earlier. If your poles have taken a beating in the past, it's a good idea to replace them before embarking on another major expedition.

Canopy strength comes not so much from heavy materials as a design that distributes the load evenly and has reinforcements at the stress points. Seams connecting pieces of uncoated fabric that are not hot-cut should either have a separate strip of fabric wrapped around the cut edges and stitched down, (a "taped" seam) or they should be flat-felled, in which the edge of each piece of fabric is folded back on itself to enclose the edge of its neighbor. The idea in both cases is to prevent a raw edge of uncoated fabric from fraying. The coating on a coated fabric serves the same purpose. Hot-cutting the fabric, which seals the edge by melting and fusing the nylon fibers, achieves the same result.

When buying a tent, check both sides of all seams for even thread tension. Load the seam gently and look for places where the seam is ready to pull apart. Then, if possible, pitch the tent. A good winter tent should pitch as taut as a sail in a stiff breeze. You shouldn't need bare hands to pitch it, though a surprising number of otherwise good tents demand that kind of dexterity. Sometimes it's possible to add a loop of webbing to the tent body to make it easier to grasp with gloved hands. If the pole diameter is large enough, you can drill a

hole through the end and tie a loop of string through it
to give you something to grab.

BUYING TENTS
The typical sporting goods store rarely carries tents
good enough to take on a serious winter camping trip.
High-quality winter tents are too expensive to appeal to
such a store's typical customer. The same is true of
most of the other gear I've been talking about. In general, you'll find the best gear in specialty mountaineering, backpacking and cross-country skiing shops. Try to
find a well-stocked shop so the sales person sells you
what you need, not what he or she happens to have on
hand.

If you can't find a specialty shop nearby, consider
mail order. Some of the best sleeping bag and tent
makers sell through mail order only. You can also find
plenty of mail-order junk. Try to get the right to return
gear unused if it doesn't seem right. When it arrives,
examine it carefully in your living room. If it looks
good, throw it in your pack and head for the hills.

· 9 ·
WHITE PALACES:
How to Build Snow Caves
and Igloos

Our fifth day on the south face of Mt. Hunter dawned stormy. Anxious about our dwindling food, we started climbing again nonetheless. Twelve hours later, with the snow still falling in blankets and darkness closing in, we began searching for a bivy site. Every possibility within view would have required hours of chopping to make it level. We got out our headlamps and kept climbing.

About 10 p.m. I saw through the mist and snow that Peter had stopped. As I joined him he said, "We might get tent platforms here!" I thrust an ice axe deep into the yielding snow and realized we might get something better: a snow cave.

Two hours later, wet and exhausted, we crawled into a snug, quiet, stormproof heaven. The cave seemed all the more luxurious in contrast to the marginal shelter of the hooped bivy sacks we'd used during the previous four nights. The stove and our body heat soon raised the temperature to near freezing. We broke into a ragged chorus of "Shelter from the Storm" and laughed for the first time in hours. We dug caves every night for the next five.

SNOW CAVES

A snow cave is the easiest snow shelter to build. That cave on Hunter was my first one. To build a simple cave, just find a hillside with five or six feet of snow on it and start digging. Keep the entrance small, and enlarge as you get deeper. To build a really snug cave, dig straight into the hillside, then up, and then begin expanding. That places the entrance lower than the floor of the cave, which helps trap warm air in the cave and keep out spindrift. You can dig a cave on flat ground, but it means moving more snow because you have to dig down first. Hillsides also form better cave sites because you can throw the snow straight out the entrance,

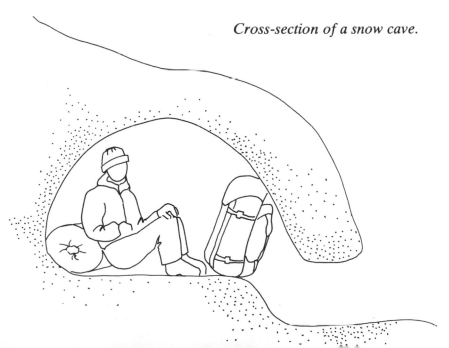

Cross-section of a snow cave.

then shovel it down the slope rather than lifting each shovelful to the surface. Poke an ice axe or ski pole through the roof for ventilation, then close the entrance once you're inside with packs or a large block of snow.

If the snow is wet, so that it sticks together well, you can ease the task of caving on flat ground by making a pile of snow several feet high and a couple of feet larger in diameter than the finished shelter you want. Pack it down well, then let it sit for 20 minutes. Shoveling snow into a pile breaks flakes into smaller particles which quickly begin to bond together if given a few minutes. When the snow has set, dig down to the side of the pile, tunnel underneath, then hollow out the pile to form a hybrid snow cave/igloo.

Be sure to avoid potential avalanche paths when siting a snow shelter. Several years ago 20 members of the Colorado Mountain Club were digging practice snow caves beneath a small hill, well below timberline. A slab avalanche over four feet thick buried two of them. Only one survived. Any open, treeless slope lying at an angle between 30 and 45 degrees, even if it's only a hundred feet high, should be considered dangerous unless you know enough about avalanches to prove otherwise.

Ideally, the cave entrance should face at right angles to the prevailing wind. If it faces directly into the wind, of course, snow blows in. Paradoxically, the same thing can happen if the entrance faces directly away from the wind and is sited just below a ridgecrest. We learned that the unpleasant way on Mt. Hunter. When we woke up after our eighth night, two inches of snow covered our sleeping bags and all our gear. Only later did I learn why. Ridges accelerate wind by compressing a large volume of air and forcing it through a smaller space, the region just above the ridgecrest. Similarly, a broad river may flow very slowly, but if forced into a narrow canyon, it will accelerate. When the river escapes the canyon, it slows once more. Eddies, places where the river actually flows upstream, may form where the river widens abruptly. In the same way, wind decelerates just after it passes over a ridgecrest and may double back on itself in a vortex. That vortex can carry snow into your cave.

The best snow-caving tool is a small, metal-bladed shovel—small so you can wield it in the tight confines of the growing cave, metal to more effectively carve out hard snow. In really hard snow, you'll need an ice axe.

Be sure to keep your cave well-ventilated. Body and stove heat can glaze the cave's interior with ice, effectively sealing it. Poke an ice axe or ski pole shaft through the roof if drifting snow threatens to block the entrance. Work the axe or ski pole in circles until you have a hole three inches in diameter. It's a good idea to bring one shovel inside, so you can dig yourself out more easily if the entrance does get buried.

Digging a cave takes a lot of wet, hard work. Don all your shell gear and your overmitts, pull up your hood and seal your cuffs before diving into the hole. Creating a decent-size two-person cave requires moving about 150 cubic feet of snow. Be prepared to spend at least an hour in soft snow, two or three in hard snow. If you hit genuine ice near the surface, try another site. You can only remove about one cubic inch of hard ice with one blow of an ice axe. To remove one cubic foot takes about 1,700 blows.

IGLOOS

Fortunately, if the snow is hard enough to make snow-caving difficult (and if you have a snow saw), you can build alternative shelters out of snow blocks. Snow saws are simply coarse-toothed saws, usually one and a half to two feet long. Pruning saws work quite well, although the sharp teeth tend to chew up your pack. Snow saws specifically designed for the task usually have blunter teeth.

To build the simplest snow-block shelter, dig a trench three feet wide and six or seven feet long. Then place big snow blocks on edge on either side and lean them up against each other like an A-frame tent. Cutting your snow blocks from a hole that will become your trench saves energy. These shelters usually accommodate only one person because blocks big enough to span a two-person trench often won't hold together.

The first-class snow-block shelter is the igloo. The procedure for building one may sound complicated, but it's actually easier than it looks. Three friends and I completed our first igloo on our first try at our base-camp below Mt. Huntington. We neglected only one problem: the potency of the sun in the Alaska Range in July. When clear skies finally returned after a week of storm, the igloo promptly collapsed.

Good snow for blocks should be firm enough that your boot heel barely dents it as you walk. Start by cutting a group of blocks of identical size, each roughly two feet long, 18 inches high and six inches thick. Arrange them in a circle, leaning inward slightly. As you place each block, bevel the vertical side so it fits flush against its neighbor. When you have one row complete, take your snow saw and turn your block wall into a circular ramp that starts at ground level and curves around to one block height after a full circle. The beginning and end of the ramp will be adjacent to each other. Now begin laying on blocks again, starting at the ramp's thinnest point. You will be building a rising spi-

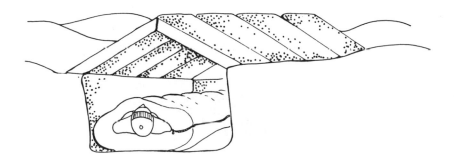

This type of snow shelter consists of a trench covered with snow blocks leaning against each other.

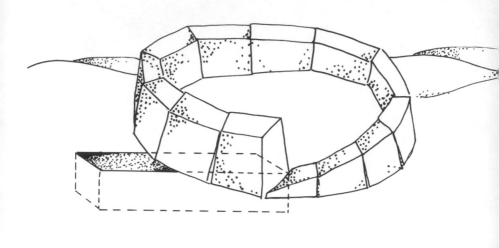

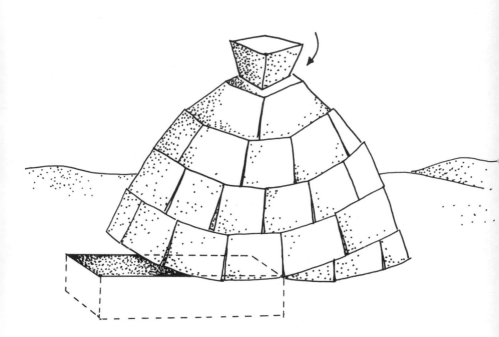

Top: *After laying down the first row of snow blocks, use a snow saw to create the "ramp" shown here. The trench jutting out to the left can be the quarry for some of the blocks. When complete, the trench serves as the entranceway.* Bottom: *To complete the igloo, shape a capstone block to fit and set it gently into place.*

ral of blocks. Each new block is supported both by the block beneath and by its neighbor to one side. It's the support of the block to one side that allows you to lean each row of blocks in a little further. One member of your group should stay inside to support and trim each block while the next is laid in place. The final block, the keystone, is trimmed to be wedge-shaped and then slipped carefully into a matching hole. Chink the holes with soft snow or bits of broken blocks, saw out a door and you've got a sturdy home. If you want to do it fancy, cut your blocks from a trench, then lay your first circle of blocks so the end of the trench is three feet inside the igloo. The trench then becomes the entrance. Since it's lower than the igloo floor, the igloo stays a bit warmer, just like in a snowcave. If you want to keep your igloo tidy but don't want to go outside to relieve yourself in the middle of the night, bring in an extra snow block to use as a chamber pot, then dispose of it outside in the morning.

SNOW WALLS

If you enjoy the light admitted by a tent, and its warmth when the sun shines, but need the shelter of an igloo, consider using snow to deflect wind from your tent. The simplest but least effective system is to dig a huge hole and pitch your tent inside. Although the walls of the hole effectively stop wind, wind-carried snow will quickly bury your tent. Wind even in the absence of precipitation can move a tremendous amount of snow. Drifts can deepen at the rate of one and a half feet per hour. However, 90 percent of that wind-carried snow is confined to the first 18 inches above the surface. To stop drifting snow from burying you immediately, build a block wall around your pit at least 18 inches high, preferably higher. It saves work to take the blocks from the area that will become the pit. You'll still get some snow piling up on your tent in high winds because the wind will decelerate slightly as it passes over the top of the wall, just as it does when passing over a ridge crest. Slower winds hold less snow, which means some snow drops out of the gale and fills in the gap between the wall and the tent.

Block walls must be substantial to withstand severe winds. Wind's first weapon is erosion. Wind whistling through the chinks in a wall gradually wears away the corners of each block. Eventually you have a stack of cannonballs instead of bricks, which can collapse with the next gust. I've seen tent poles break when struck by falling walls. You can strengthen both snow walls and igloos by chinking the cracks and by pouring water over the junctions between blocks to freeze them together.

Really vicious winds can simply knock walls down. In 1984 I built a wall around my tent at 17,200 feet on McKinley using blocks two feet long and a foot square. Moving each block made me gasp. After the wind knocked the wall down twice, I finally rebuilt it using the blocks so that their long axis was perpendicular to the wall. That made the wall two feet thick. At last I had a "Maginot Line" that could withstand the wind's punishment. It took so much work to build that I was joking about auctioning it off when I left. But I slept soundly once it was done, which made it worth every ounce of energy.

· 10 ·
CARE OF THE HUMAN ANIMAL:
Prevention, Recognition and Treatment of Hypothermia and Frostbite

I knew that moving our camp from 14,300 feet to 17,200 feet on McKinley would stress my clients severely. But I didn't know how close one client—Marilyn—would come to serious trouble.

The day began with a 2,000-foot climb up an ever-steepening snow wall. Everyone, particularly Marilyn, was tired at the top, and we still had to gain 1,000 feet of elevation up an exposed ridge. The wind began picking up.

We pulled into camp about 5 p.m. The increasing wind and dropping temperatures forced everyone, including Marilyn, to rummage through their packs for more warm clothes. We began digging a giant snow cave in case McKinley's weather machine really ran amok.

Marilyn chipped in for a while, then, without my noticing it, slipped into the tent we had erected as temporary shelter for people while they added extra clothing. Half an hour later, a client emerged from the tent after pulling on another layer and said to me, "Marilyn seems to be getting very cold. Maybe you should check on her."

I crawled into the tent immediately and called her name. She only moaned in reply and suddenly I got worried. Fortunately, two of my clients had bags that zipped together. I borrowed those and zipped them into one giant bag. Then an assistant and I shed some of our outer layers, removed some of Marilyn's outer layers, put her in the double sleeping bag and crawled in with her. We wanted our warmth to reach her without obstruction by lots of insulation, which keeps heat out as effectively as it keeps heat in. Meanwhile, another assistant managed to get a stove going outside in the wind. Soon he brought in a bottle of piping hot water and, shortly thereafter, another. We laid them on Marilyn's stomach.

To my relief, she began to warm up within an hour. When she was fully conscious, we gave her sweet drinks, then soup, then solid food. Soon she was able to move into her own sleeping bag. Although bouts of shivering continued to attack her occasionally for several more hours, the danger was clearly over.

For a while, Marilyn had been seriously hypothermic. Had the dangerous drop in her core body temperature continued untreated, it could have been fatal. Below a core temperature of about 94 degrees, people can become incapable of rewarming themselves no matter how much insulation is added. Some form of external heat must be provided.

Almost every possible factor had combined to predispose Marilyn to hypothermia. She hadn't been eating enough for days, in part because altitude and exertion had made her stomach queasy. Not eating enough meant she had little fuel to burn to generate warmth. Her low blood sugar, a consequence of eating too little, decreased her ability to shiver, one of the body's main mechanisms for generating heat. She also hadn't been drinking enough and was almost certainly dehydrated. Dehydration aggravates the effects of hypothermia.

Marilyn's inadequate food intake had slimmed her already trim build still further. Thin people have less insulating fat, so they cool faster. She also had the disadvantage of being fairly short. Small people have a greater surface area in proportion to their volume than big people, which means faster heat loss. Infants are the most susceptible of all in this regard, particularly because they can't shiver effectively.

Marilyn was also not very fit, which meant the hard work of climbing to 17,200 feet exhausted her even more than the rest of us. Exhaustion also reduced her ability to shiver. The amount of oxygen consumed by an active, mildly hypothermic individual is much larger than the amount consumed by the same person when warm. That oxygen consumption reflects the greater energy cost of working with cold muscles. Marilyn probably started to get chilled even before she reached camp. Had she stayed warm during the day, she would probably have arrived less exhausted. Hypothermia and exhaustion link arms, each goading the other on.

The final contributing factor to Marilyn's hypothermia was the high altitude. The air's lower oxygen content probably caused her to become somewhat hypoxic, a condition in which insufficient oxygen reaches the body's tissues. Hypoxia also decreases shivering.

Simply being a woman probably didn't affect her chances of hypothermia. Although women, on average, are smaller than men and have a slightly lower basal metabolism (amount of heat produced at rest), their higher level of insulating body fat evens the score. Women in cold-water immersion tests cool at very nearly the same rate as men. Other factors besides sex play a larger role in who is most susceptible.

Fortunately, Marilyn didn't make the mistake of drinking alcohol or taking tranquilizers or sleeping pills. These drugs decrease shivering and help predispose a person to hypothermia. She also wasn't injured. Injuries, by causing shock, can make victims more susceptible to rapid cooling.

Given everything Marilyn had working against her, I didn't have much hope she'd make the summit. Although reaching the top from 17,200 feet takes only a day, the cold, wind and even higher altitude conspire to make that day even more grueling than the day getting to 17,200. We hunkered down in our tents and began waiting for the weather to clear.

So far in this book, I've talked mostly about using gear intelligently to slow the rate of heat loss from your body. Now it's time to talk about the other side of the equation: how your body generates heat and what you can do to help it.

Heat generation must balance heat loss or you will soon be either sweating or shivering. Your body generates heat by breaking down food molecules. When you're inactive, nearly all the energy derived from the metabolism of food eventually appears as heat. When you're active, roughly 20 percent of the chemical energy in food is used to move the working muscles. The rest becomes heat.

A few animals, such as rats, can acclimatize to cold by learning to burn food faster, thereby generating more heat, without constant shivering and without working the large muscles. Studies of men winterizing in Antarctica, however, have found only debatable evidence that people raised in temperate climates can achieve an equivalent kind of cold acclimatization, even given several months. There is some evidence that cold-acclimatized Antarctic residents experience a more rapid and severe vasoconstriction when exposed to cold than they did before going to Antarctica, but this is a means of restricting heat loss, not increasing heat production. Intense vasoconstriction predisposes your hands and feet to frostbite, besides being quite uncomfortable, so it cannot be an adequate defense against the cold.

Hands, by contrast, acclimatize to cold quite well. Even if you're not used to cold, you will experience something called "cold-induced vasodilatation" when your hands are chilled. Every seven to 15 minutes, the vasoconstriction in your hands will be replaced briefly by vasodilatation. This "hunting reaction" causes your hands to feel warmer and less stiff. Expose your hands to cold long enough and often enough, and the periods of vasodilatation will become more frequent, more intense, and longer-lived. Cold-induced vasodilatation can affect the whole body as well, particularly in people immersed in cold water. The sensation is said to be like "basking in the cold." Unfortunately, the net effect of whole-body vasodilatation is simply faster core heat loss.

A few groups of people have developed an amazing

tolerance for cold. Charles Darwin reported seeing native women in Tierra del Fuego nursing their babies while snow fell on their bare breasts, apparently oblivious to the cold. Australian aborigines can remain asleep when their core temperature begins to drop, unlike non-aborigines, who almost invariably wake up shivering miserably. No one fully understands the mechanisms of this cold adaptation, despite a good deal of research. For most people, cold acclimatization is something to fantasize about as we reach for another heavy sweater. Without clothing, most people are only comfortable when the temperature exceeds 80 degrees F.

Unless you're an aborigine or Tierra del Fuegan, shivering is the only mechanism by which you can consistently generate more heat when you're inactive. Intense shivering increases heat production to five times the resting rate. Strenuous exercise, however, increases it even more, and is much more useful if you use that energy to build an igloo or hike like a runaway locomotive to escape a cold situation you can't deal with. Even the most vigorous exercise, however, cannot compensate for heat lost in severe cold unless you have adequate clothing.

As you can see, you can't do much short of shivering or performing jumping jacks to increase your body's heat production. But you can do a lot to keep heat production from falling. To start with, that means eating enough food.

The amount of food required when you're inactive, whether in the cold or not, is pretty constant. That's because your basal metabolic rate doesn't rise significantly in cold weather. However, movement in the cold does cost more energy than moving in a warm climate, primarily because you're working against the resistance of all that heavy clothing. You're also constantly picking up heavy boots and frequently picking up skis or snowshoes as well. In addition, you're fighting the resistance of the snow. Two studies in Antarctica concluded that outdoor work cost twice as much energy there as it would have in a temperate climate.

Eating a meal increases metabolism by about ten percent for the next four to six hours. You should eat at least that often, and many experienced winter travelers stop every two hours or so. Never let yourself get ravenous. Munching frequently has the added advantage that your nibble-stops need not be long. That helps prevent the chilling that happens so easily if you stop for a long time and neglect to add clothing. A friend calls this the "shark theory of staying warm." Most sharks must swim constantly to keep water flowing through their gills, which is what enables them to breathe. In a similar way, humans in severe cold stay warm most easily by moving nearly constantly when they're not in their sleeping bags.

Carbohydrates and protein give about 110 calories per ounce, fats about 250. It's tempting to increase the fat content of your diet if you're concerned about the total number of calories you can carry on your back, but a fatty diet quickly becomes unpalatable. It's also bad for your health over the long term. Carbohydrates digest faster than fats, so if you need some energy quickly, go for them. On a long trip, you'll need to balance all three major kinds of food to get adequate amounts of vitamins and minerals as well as calories.

Drinking inadequate amounts of water doesn't directly cause hypothermia, but it greatly aggravates its effects. To understand that, you need to understand how hypothermia alone affects the circulatory system.

Cold acts directly to increase blood's viscosity. Cold also causes some of the blood's water content to leave the circulatory system and become trapped in the body's other tissues, which increases viscosity further. Finally, cold causes the spleen to contract, which increases the number of red blood cells in circulation, again increasing viscosity.

Cold also causes the overall volume of blood to drop through a phenomenon called "cold diuresis." When you start to get cold, the blood vessels near the skin squeeze down. All that blood is shunted to the core, where the kidneys perceive it as surplus. The kidneys then extract the water, which passes out as urine.

The result of all these different effects of cold on circulation is a low volume of thick, hard-to-pump blood being pushed around by a heart that itself becomes weaker as its temperature drops. Since blood carries oxygen, that means the body's cells receive less oxygen. Oxygen is critical to their normal functioning.

Dehydration compounds hypothermia's dangerous effects on circulation by further reducing blood volume. The blood thickens still more, circulates more slowly and provides even less oxygen to the oxygen-starved cells.

Fluid losses go way up in the cold even if you avoid cold diuresis by staying warm. Cold air holds little water vapor compared to warm air, even when the cold air's relative humidity is 100 percent. Cold air is always dry.

When you inhale that dry air, you warm it through contact with your nose, mouth, throat and lungs. The warmed air absorbs moisture from the moist linings of your respiratory system. The result is that fluid losses from your lungs go up. Given the same level of activity, you always need to drink more in the cold than you do at moderate temperatures.

High fluid losses, combined with the serious consequences of failing to drink enough, mean you must drink a lot to stay healthy in the cold. Unfortunately, your body won't tell you accurately how much you need to drink. It will tell you when you're thirsty, but simply putting something in your mouth decreases the sensation of thirst for about 15 minutes. People tend to drink a cup when they need a couple of quarts. The easiest way to tell if you're drinking enough is to monitor the color of your urine. Passing large quantities of clear urine (about one and a half quarts every 24 hours, if you care to measure) is a good sign you're drinking enough. A volume of less than half a quart signals that you're getting dehydrated.

To pass a quart and a half of urine, you may need to drink anywhere from half a gallon to a gallon and a half of water every day, depending on the degree of cold, your level of exertion, and, most importantly, your altitude. Mountaineers at high altitude may need as much as a gallon a day just to humidify the air they breathe, according to James Wilkerson, co-author of *Hypothermia, Frostbite and Other Cold Injuries*. Sweat, urine output and insensible perspiration can easily account for another two quarts.

High altitude also affects susceptibility to hypothermia in other ways. Water evaporating from lung and throat linings extracts heat just like it does when it evaporates from your skin. Panting in the thin air increases heat loss as well as fluid loss. Loss of appetite and insensitivity to thirst also commonly afflict people at altitude. To defend yourself when you go high, you must consciously force yourself to eat and drink enough. If your dipstick reads low, you don't ask your car whether it's thirsty before giving it a quart of oil. In the same way, you don't ask your body whether it's thirsty or hungry when you pull into a high-altitude camp. You just start drinking and munching.

A reliable stove to melt water from snow is a crucial piece of cold-weather equipment. Backpacking stoves are available that burn white gas (more properly called naptha), kerosene and butane. The easiest to find are the white-gas and butane models. In general, white-gas stoves that come with a pump for increasing the pressure forcing vaporized gasoline through the nozzle provide the highest heat output in the cold. They can be temperamental, however, so look for a model that can be stripped down, cleaned and repaired in the field with simple tools and spare parts. Butane stoves, because they're simpler, are often more reliable. The problem with butane is that it won't vaporize at temperatures much below freezing. A stock butane stove is nearly worthless in the cold. I still remember slogging back to my car after my first multi-day winter climbs, grossly dehydrated, my voice a croak, cursing the useless butane stove in my pack.

Cold butane stoves work better at high altitude than at low because the lower atmospheric pressure means a lower boiling point for butane, just as it does for water. More butane vaporizes per minute, and so the stove burns hotter.

Armed with knowledge, experience and the right equipment, you need not fear hypothermia. But the odds are decent that some day a member of your group, or someone you meet in the backcountry, will become hypothermic. Here's how to recognize the problem and correct it.

For the purpose of discussion, let's divide hypothermia into two types, mild and profound, although the progression of symptoms isn't broken into sharply defined stages. Mild hypothermia begins with cold toes and fingers and a general sensation of feeling chilly. Victims are usually shivering, though that may not be apparent when they're moving. People who are shivering clearly need to get their heat equations back in balance.

Mild hypothermia causes other problems, too. As muscles cool, they become progressively stiffer, weaker and less coordinated. Clumsiness usually strikes hands first. More severely hypothermic victims frequently have trouble walking on rough terrain. Their speech may become slurred. They usually lose interest in whatever goal they were originally pursuing and become interested only in getting warm.

In profound hypothermia, muscular deterioration becomes more pronounced, until victims may no longer be able to stand. Shivering usually stops at a core temperature of 90 to 92, although individuals vary widely in this. Apathy takes over completely. Profoundly hypothermic people frequently fail to do all they can to keep themselves from getting colder. They leave jacket zippers unzipped, or fail to get all the way into their sleeping bag, or forget to put on hats and gloves. An even more serious confusion follows. Victims of profound hypothermia sometimes fling off their clothes. As cooling continues, victims become comatose. Heart and respiration rates slow and may become practically imperceptible. Eventually, heart failure causes death.

Treatment for mild hypothermia is simple: rewarm the victim immediately by any convenient means! First, stop the heat loss. If the victim has fallen into a lake or stream, take off the wet clothing and put on dry. If you don't have dry clothing, at least wring out the wet clothing thoroughly, then put it back on. Put something waterproof and not breathable on top: raingear, a plastic groundsheet, a large plastic garbage bag with holes for the head and arms. If the victim is dry, add more clothing, or get him or her into a sleeping bag. People with mild hypothermia will eventually rewarm by themselves if they put on lots of dry insulation. To speed the process, add heat. Build a fire or place hot water bottles along the sides of the chest and where the thighs meet the torso. Better yet, get a warm person into the same sleeping bag with the cold one. Zip two bags together if there isn't room inside one.

Hot liquids make a victim feel better and combat dehydration, although they have less actual warming effect on the core than you might think. A pint of hot liquid in the stomach of a 150-pound man provides enough heat to raise the man's temperature by less than half a degree. The feeling of warmth comes because hot liquid flowing down the throat warms the blood flowing to the brain. The brain, thinking the body is too warm, responds by dilating blood vessels in the skin. That makes the victim feel better. Alcohol also causes vasodilatation, but its other effects are harmful. Alcohol decreases shivering, acts as a diuretic, accelerating dehydration, and impairs judgment. Alcohol shouldn't be given to any hypothermia victim. Shivering can stop if skin temperature goes up, but the core remains cool, so don't leave the fire or shed clothing too soon.

Profound hypothermia is a truly life-threatening emergency. Death can occur through heart failure even after rewarming begins through a complex phenomenon called rewarming shock. Cold has so weakened the heart in profoundly hypothermic victims that simply handling them roughly can cause ventricular fibrillation, in which the heart muscle fibers begin contracting randomly rather than in synch. Death follows within minutes. Victims should not try to help themselves. Literally any movement can cause sudden death. If you can prevent a victim from cooling any further and a professional rescue team can be summoned within a few hours, your best option is turning the rescue over to them. If that's not possible, very gentle, slow rewarming of the torso only, with absolutely no rough handling, may save a life.

Frostbite is the other major danger that cold poses for winter travelers. The same care of your body that will help prevent hypothermia goes a long ways towards preventing frostbite. Keep your core temperature up, and your extremities will stay a lot warmer. That means eating enough as well as wearing appropriate clothing. Wind chills bare skin much faster than calm air at the same temperature. Beware of removing your gloves in high wind. Dehydration contributes to frostbite because the drop in blood volume causes the blood to thicken, as I explained earlier. Viscous blood flows more slowly, so it cools further before reaching your extremities, providing less warmth. High altitude also increases the risk of frostbite because altitude decreases cold-induced vasodilatation, even after acclimatization to the thin air.

Tight boots and gloves constrict circulation and greatly increase the risk of frostbite. Stuffing too many socks or liner gloves into your boots or mitten shells, or lacing your bootlaces or crampon straps too tightly, has the same effect. Avoid nicotine, which is a vasoconstrictor. Don't handle metal objects with bare hands. Metal conducts heat more than a thousand times faster than air. On a nice windless, zero-degree day, you may be able to leave off your gloves for several minutes without great discomfort. Grab the metal head of your ice axe, however, and you can have a nasty case of frostbite within seconds.

Extremely cold liquids are equally dangerous because contact between the cold liquid and your skin is so intimate and evaporation is so fast. Gasoline, kerosene and alcohol all freeze at a much lower temperature than water. Spilling very cold gasoline or kerosene on your skin will cause serious injury. If you leave a bottle of high-proof booze outside your tent on a cold night, then take a swig in the morning, you can freeze the inside of your mouth.

Frostbite's first warning is usually pain, although not for all people. Numbness follows, though again not for everyone. Once the pain stops, it's tempting to think the problem has vanished. Unfortunately, it may only be getting worse. I had no inkling my fingers were becoming frostbitten on Mt. Hunter until I removed my gloves to tighten a crampon strap.

Frostbitten skin usually appears white and hard. If a large area, such as an entire hand or foot, is frozen, the skin may appear purplish because the blood flowing sluggishly in the vessels has clotted. That kind of discoloration usually means much or all of the frostbitten part will be lost.

Frostbite damages tissue in two ways. First, ice crystals form between the cells. These ice crystals extract water from the cells as they grow, damaging the cells physically and chemically. Ice crystals can form within cells, but only if freezing occurs much faster than is normal in frostbite injuries.

More damaging to the tissue than ice crystal formation is the obstruction of the tissue's blood supply. As I mentioned earlier, cold affects the cells lining the capillaries in such a way that they allow the water portion of the blood to leak out into the tissues. The blood that does continue to flow thickens and eventually clots. No oxygen reaches the tissues, leading to permanent damage.

The best treatment for frostbite is rapid rewarming in a water bath between 100 and 108 degrees. In no case should the water temperature exceed 112 degrees. That means it should not feel hot to an uninjured hand. To keep the pot or tub warm, remove the injured limb and add hot water. Stir thoroughly, check the temperature, then put the limb back in. Never use an open flame under the pot to keep it warm. The injured area, while still numb, could come in contact with the bottom of the pot and be severely burned without the victim's knowledge.

Continue thawing for 20 to 40 minutes, or until the frozen part flushes pink. Taking one or two aspirin while the part is still frozen and every six hours thereafter may help increase circulation through aspirin's anti-clotting properties.

Slow thawing at a lower temperature usually leads to greater tissue loss. Thawing at higher temperatures, such as in front of a fire or over the exhaust of a snowmobile, is a sure prescription for disastrous loss of flesh. The victim cannot feel the heat, which often means that a burn is added on top of the frostbite.

Once thawed, a frostbitten extremity will be very delicate. Frequently large blisters form, which should be protected. Infection, always a threat, is even more likely if the blisters are broken. Combating infection is hard because the circulation is so poor that antibiotics don't get where they're needed. Don't thaw frostbitten feet if the victim must walk to get out. You'll cause much less damage walking on frozen feet than on thawed feet. The swelling that accompanies thawing will probably mean the victim's boots will no longer fit. The pain may prohibit walking in any case.

Never thaw a part when there is significant danger of it refreezing. Refreezing causes much more damage than allowing a part to remain frozen for a few hours, even a day or so. Also, don't thaw just the frostbitten portion of a hypothermic patient without treating the hypothermia as well. Once the limb is thawed, a good blood supply is crucial. The vasoconstriction caused by the hypothermia will deprive the thawed limb of blood when it needs it most.

Never rub a frostbitten limb with snow. The idea that such cold therapy was helpful came from the observations of Baron Larrey, surgeon-general of Napoleon's army during the retreat from Moscow in the winter of 1812–1813. Frostbite was epidemic. The soldiers thawed their limbs each night over a fire, then refroze them during the day. Infection ran wild, and many soldiers lost fingers, toes and more. Larrey concluded that heat of any kind caused infection. He was certainly right about the harmful effects of excessive heat, but wrong that the solution was to treat cold injury with cold.

Actual freezing isn't necessary to damage limbs from cold. Trench foot occurs when the victim's feet stay wet and cold continuously for days on end. Common symptoms include swelling, redness, blisters and nerve damage. The damage can be permanent, and is sometimes so severe, if the exposure to cold continues for weeks, that amputation becomes necessary. Although trench foot is potentially very serious, prevention is simple. Remove your boots every night, dry your feet, warm them if they're cold and give them a good massage to promote circulation. Treatment in most cases is equally simple: keep your feet dry, warm and elevated (to reduce the swelling).

Chilblains can become a minor annoyance to people who expose bare skin to wet, windy conditions with the temperature above freezing. The skin becomes red, rough, dry and itchy and sometimes cracked. Most people with chilblains would say they have badly chapped skin. Put on a moisturizing lotion, keep the skin warm and it will soon be fine.

Preventing hypothermia and frostbite takes savvy, conscientious care of your body and adequate gear. Going a little further and becoming truly at ease in winter weather takes a little more of each of those ingredients. I think you'll find that the rewards for the effort are high indeed. Marilyn did hardly anything right on her way up McKinley to 17,200 feet. But she learned, and she had the courage to keep trying even after a serious bout of hypothermia. Five days after reaching 17,200 feet, she stood on the summit, warm, relaxed and about as happy a person as I've ever seen. For her, learning to deal with the cold made a longstanding dream possible. Climbing McKinley was one of the most satisfying adventures of her life. I hope that learning to live comfortably in the cold will bring as many rewards to you.

APPENDIX
My Personal Layering System

This is the clothing I bring for a typical winter outing. The amount you will need may vary.

For a one-day ski tour when the temperature is above zero:

thin polypro or polyester long underwear top
thick polypro or polyester long underwear top with
   polypro or polyester balaclava sewn on
quilted polypro or polyester zip-front jacket
shell jacket
ski hat
pile balaclava
sun glasses
goggles
Neoprene face mask
pile bibs
shell pants
one pair of liner gloves
one pair of heavy wool gloves
two pair of insulated overmitts
one pair of liner socks
one pair of heavy wool socks
double ski-touring boots
uninsulated gaiters

If I'm going on an overnight ski tour, I add:

one pair of long underwear bottoms
down jacket

If I'm going on a high-altitude expedition, I add:

two pair of long underwear bottoms, one thin, one
   thick
down jacket
vapor-barrier shirt
heavy pile jacket
a second pair of liner gloves
one pair of heavy mittens
vapor-barrier socks
a second pair of liner socks
a second pair of heavy wool socks
foam insoles between the inner and outer boots
supergaiters with extra foam and pile insulation

Index